PHalarope books are designed specifically for the amateur naturalist. These volumes represent excellence in natural history publishing. Most books in the PHalarope series are based on a nature course or program at the college or adult education level or are sponsored by a museum or nature center. Each PHalarope book reflects the author's teaching ability as well as writing ability. Among the books:

The Amateur Naturalist's Handbook
Vinson Brown

At the Sea's Edge: An Introduction to Coastal Oceanography for the Amateur Naturalist
William T. Fox/illustrated by Clare Walker Leslie

Biography of a Planet: Geology, Astronomy, and the Evolution of Life on Earth
Chet Raymo

Botany in the Field: An Introduction to Plant Communities for the Amateur Naturalist
Jane Scott

The Crust of Our Earth: An Armchair Traveler's Guide to the New Geology
Chet Raymo

A Field Guide to the Familiar: Learning to Observe the Natural World
Gale Lawrence

The Fossil Collector's Handbook: A Paleontology Field Guide
James Reid Macdonald

Nature in the Northwest: An Introduction to the Natural History and Ecology of the Northwestern United States from the Rockies to the Pacific
Susan Schwartz/photographs by Bob and Ira Spring

Nature with Children of All Ages: Activities and Adventures for Exploring, Learning, and Enjoying the World Around Us
Edith A. Sisson, the Massachusetts Audubon Society

The Plant Observer's Guidebook: A Field Botany Manual for the Amateur Naturalist
Charles E. Roth

The Seaside Naturalist: A Guide to Nature Study at the Seashore
Deborah A. Coulombe

365 Starry Nights: An Introduction to Astronomy for Every Night of the Year
Chet Raymo

A FIELD MANUAL FOR THE AMATEUR GEOLOGIST

TOOLS AND ACTIVITIES FOR EXPLORING OUR PLANET

Alan M. Cvancara

PRENTICE HALL PRESS • NEW YORK

Published in 1986 by Prentice Hall Press
A Division of Simon & Schuster, Inc.
Gulf + Western Building
One Gulf + Western Plaza
New York, NY 10023

Originally published by Prentice-Hall, Inc.

Library of Congress Cataloging-in-Publication Data

Cvancara, Alan M.
A field manual for the amateur geologist.

(Phalarope books)
Bibliography: p.
Includes index.
1. Geology—Guide books. I. Title.
QE45.C83 1985 550'.72 84-11596
ISBN 0-13-316522-1 (pbk.)

Manufactured in the United States of America

10 9 8 7 6 5 4 3 2 1

First Prentice Hall Press Edition

To Ella: wife, mother, copy editor, critic, typist, listener, provider of a suitable environment for one who desired to try his hand at writing, and trusted friend.

CONTENTS

Contemplating the past

Minerals, rocks, and fossils: the stuff of geology

How to do geology

Appendixes

PREFACE

This book is for a lover of the outdoors, a hiker, an amateur naturalist, or anyone who wishes to learn some geology "on the run." So, I hope this field manual will be particularly useful to you, the traveler—whether you are in a car, a train, a plane, on a bike, or on foot. Learning geology while traveling is learning geology best, with least effort. A week in the field with this manual is worth at least a few weeks in the classroom.

The first part concentrates on identifying landforms from a moving conveyance and explains how they form and change with time. The second part emphasizes geology from a chronological distance, not a physical one. This part is best digested when you are stationary: by a campfire at day's end, or waiting out a rain- or snowstorm.

The third and fourth parts encourage you to become at times a sojourner in order to experience geology more intimately. "Minerals, Rocks, and Fossils: The Stuff of Geology" is an aid to the identification of the most common earth materials and their features. "How to Do Geology" explains how a geologist operates and how you can accomplish several geological activities. Here, you will learn some of the tools of the trade.

Whether you wish to learn some geology—with relative ease—from a distance or close-up, may this book help you not only to learn about your physical environment but also to enjoy it.

Now a couple of points regarding procedure. I have included additional suggested readings at the ends of most chapters. These were selected for their relatively nontechnical writing. Technical terms in this book stand out in boldface or italics where defined; I consider those in boldface more significant.

A Strategy for Geologic Excursions

If you are new to geology, skim this book—especially the first part—before departing on a trip. And, if you have time to plan a route, acquire maps other than general highway maps. I recommend the regional geological highway maps listed in Appendix C. Should you intend to cover less ground in more detail, selected topographic maps (see Chapter 1) will prove useful. Assemble a few items of equipment. For mineral, rock, or fossil collecting, take along a geologist's or bricklayer's hammer, hand lens, acid bottle, and collecting supplies (see Chapter 21), most of which can be fitted in a knapsack. Binoculars and camera should be included, and, of course, a notebook for recording observations and materials collected. Visit parks and museums (see Appendix A) in and out of parks. Stop in at state geological survey offices (see Appendix B), or inquire of them beforehand about geological features to visit and state geological maps and publications. With continued exposure to geological information, you may find it desirable to carry a small geological dictionary. A useful one, published in a pocket-sized, paperback format, is the American Geological Institute's *Dictionary of Geological Terms.* Or, perhaps, Alec Watt's *Barnes and Noble Thesaurus of Geology* in which many features are illustrated in color.

As you travel, observe carefully, and write down the observations. Analyze, question, and interpret. This is how geologists learn. In like manner, such a strategy will enable you to learn too.

Acknowledgments

My parents, Charles and Lillian, now deceased, provided me with the means, stimulus, and encouragement to acquire a formal education—much beyond their own—that ultimately led to the writing of this book. And Mary E. Kennan of Prentice-Hall furnished editorial incentive necessary for the writing of the book. Providers of photographs and drawings are acknowledged in the figure captions. Unacknowledged illustrations are my own.

A FIELD MANUAL FOR THE AMATEUR GEOLOGIST

The lay of the land: landforms

1 INTRODUCTION TO LANDFORMS

Constructional Versus Destructional Processes

The physical world is the scene of battles. These are barely noticeable at some times, seen as minor skirmishes at others, and, occasionally, culminate in a major flare-up, as when a volcano erupts. Combatants group into two camps: **destructional processes** that tear down the land and **constructional processes** that build it up. Destructional processes include **weathering** (the in-place breakup of rocks), the downslope movement of earth materials, and **erosion** (the weathering, wearing away, and transport of earth materials by running water, ground water, wind, and glacial ice). Solar energy and gravity drive these processes to accomplish their geologic work. Solar energy evaporates sea and lake water that later falls and collects into streams or precipitates as snow and eventually compacts to glacial ice. Both streams and glaciers then erode, aided by the downward force of gravity. Solar energy also generates wind that erodes land directly or through the action of waves and currents along shorelines. Besides driving streams and glaciers, gravity impels earth materials down slopes directly; when caused by the pull of the sun and the moon on the earth, it creates tides that allow waves and currents to erode at more than one level. And tidal currents themselves may be capable of erosion.

Constructional processes are **volcanism**—including the formation of volcanoes and lava flows; earth movements—especially those that raise the land to provide, with volcanism, more fuel for destruction; and **deposition,** or the laying down of cool, nonmolten material by running water, ground water, wind, and glacial ice. So, you see, these agents construct as well as destroy.

Both destructional and constructional processes mold the myriad of **landforms** that constitute the surface features of the earth and are the subjects of Chapters 2 to 10. Erosional landforms, those resulting from the effects of wearing down, are the most conspicuous and owe their identity primarily to the relative resistance of the materials eroded. Depositional landforms nearly always occur in low-lying areas, below the source of the materials that constitute them.

Topographic Maps and Aerial Photographs as Aids in Identifying Landforms

Landforms are characterized by their setting or surroundings, ground plan and profiles, internal make-up, and surface features. The first two features are particularly well displayed on topographic maps and aerial photographs. Use these tools to help you identify landforms, especially from the air.

Topographic maps (Figure 1–1) show topography—the lay of the land or the configuration of the earth's surface—by **contours** or lines connecting points having the same elevation. They wrap horizontally around hills, curve outward around ridges, and bend up valleys before crossing streams. The main point to remember about them is this: Closely spaced contours mean steep slopes; widely spaced contours signify gentle slopes. A vertical cliff on a topographic map appears as many contours drawn on top of one another. Generally, every fifth contour is darker and wider and labeled with its elevation above (or below) mean sea level. If contours seem unclear to you, imagine a volcanic island in the Pacific. Water level corresponds to the zero contour circumscribing the island. If the sea level drops 20 feet (6 m), remains for a time for a beach to form, and drops twice more in the same way, two beach lines will encircle the island above present sea level. The beach lines can represent contours at 20 feet (6 m) and 40 feet (12 m) above sea level.

To find the height of a landform, you need to know the **contour interval** or difference in elevation between adjacent contours; it is given at the base of a map. So, if a hill on a map has five contours wrapped around it, and the contour interval is 20 feet (6 m), the hill is at least 80 feet (24 m) high. To be more precise, suppose the lowest encircling contour has an elevation of 2000 feet. A spot elevation at the top of the hill reads 2093 feet; therefore, the true height of the hill is 2093–2000 or 93 feet (28 m). Encircling contours may also depict enclosed depressions if they are set with short lines at right angles to them, pointing inward, and downslope such as would depict a volcanic crater. Counting contours here would give you the crater's depth. Contours can be thought of, too, as portraying the *relief* of a region, or the difference in elevation between the high and low points. A region may be one of low or high relief, and you can pick off specific values of elevation readily. Some topographic maps also show relief by shading (Figure 1–1).

To find the ground plan dimensions of a landform, you must know the map's **scale,** or the ratio of distance on the map to that on the ground. This is given at the base of the map and is most usable in the form of a divided bar or line (see also Figure 20–3). With a strip of paper, mark off the dimensions of the landform and compare them to the scale to obtain values. Maps come in various scales: Small-scale maps show large areas smaller and in less detail, large-scale maps show small areas larger and in more detail. Generally, you will find both types useful for the same landforms, because small-scale maps show setting and patterns and large-scale maps show useful details.

You might use topographic maps before visiting a region to help spot landforms and after a visit to check on observations or measure landform dimensions. To sharpen your identification eye, you might examine compilations of maps selected for their good portrayal of landforms. Two, DeBruin's *100 Topographic Maps* and Upton's *Landforms and Topographic Maps,* are useful for this purpose.

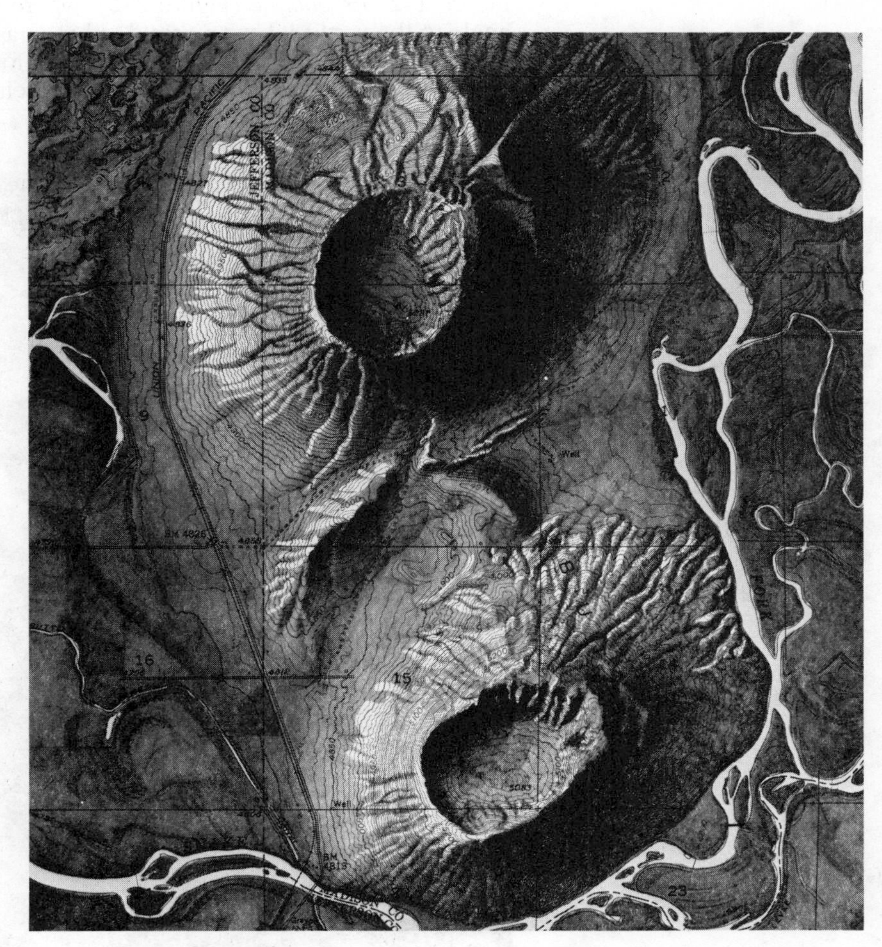

FIGURE 1–1. Topographic map of two volcanic cinder cones (Menan Buttes) near Menan, southeastern Idaho, showing relief by shading as well as by contours; the cones project about 500 feet (150 m) above the surrounding lava plain, incised by the meandering Snake River. Squares of the grid are one mile (1.6 km) on a side. (From the U. S. Geological Survey Menan Buttes Quadrangle, 1954.)

How can you examine maps of any region in the United States? The most likely places to look are libraries of universities or state geological surveys. If these or similar sources are unavailable, you can purchase the maps. Request an Index to Topographic Maps for your state of interest from the National Cartographic Information Center, U. S. Geological Survey, 507 National Center, Reston, Virginia 22092. From the index map, select your *quadrangle maps*—they cover areas bounded by lines of latitude and longitude—by name and scale. The scale depends on the area covered by the quadrangle. Order maps according to the instructions of the National Cartographic Information Center.

Aerial photographs are either *vertical* (most common—see Figure 1–2) or *oblique* (Figure 2–2), where the view is at some angle to the vertical. Lower altitude photographs, of course, show the most detail. Oblique photographs allow for rapid identification of landforms, but vertical photographs are useful for landform study and analysis.

FIGURE 1–2. Vertical aerial photograph of the area shown in Figure 1–1. Part of a lava flow is visible in the upper left, and fields are evident on the Snake River flood plain (Chapter 2). (U. S. Geological Survey photograph Idaho 7 B.)

To identify landforms on vertical photographs, look for differences in tone—the relative amount of light a feature reflects; the degree of dissection or erosion of a surface (see Chapter 2); and patterns as they relate to deformed (and eroded) rocks as well as stream drainages (see Chapter 9). Since most aerial photographs are taken in black and white, you must concentrate on the shades of gray. Black usually indicates water, but it could also reflect a lava flow. Darker grays may mean materials that retain more moisture (clayey soils, shale), lower areas, or forested regions. Lighter grays may indicate materials that retain little moisture (sandy soils, sandstone), higher areas, or grassland areas. Uniform tone represents uniform soil or rock type and moisture, as found on lake plains, or thick, flat-lying rocks. Mottled tone implies an inhomogeneity to a surface, with light, drier areas interspersed with dark, lower intervening depressions. Banded tones may reflect differences in the color of flat-lying or tilted beds as well as differences in soil or rock type or moisture availability. Certain vegetation may grow on a particular landform and give a clue to its identity. This gives you some idea how to read aerial photographs, but you must learn by using them.

Vertical photographs are taken along flight strips with about 60 percent overlap at each end; and the flight strips overlap along their sides about 30 percent. So each photographed spot appears at least twice. Take two consecutive photographs, each from a slightly different position along a flight strip—called a *stereopair*—and view them through a *stereoscope,* a gadget with two lenses on short legs.* Result: three-dimensional portrayal as with old-time stereoscopes or with special glasses at suspense-filled 3-D movies. Some individuals can see in 3-D immediately, others require practice. If the stereopair is permanently mounted (Figure 1–3), the task is easier. Adjust the space between the lenses on the stereoscope to correspond to that between your eyeballs—about 2.4 inches (6 cm). Look down on the stereopair through the lenses, your nose over the line separating the two photographs. One eye should be over the feature to be viewed in one photograph, the other over the same feature in the second photograph. The trick now is to stare *straight down,* eyes viewing in parallel paths, and attempt to focus as if looking into space. If everything works, the eyes and brain fuse the two images into one. With unmounted photographs, you must position them so that the desired feature on each is separated by a distance about equal to that between your eyes. Many persons, with experience, can see the three-dimensional portrayal without a stereoscope. Remember that vertical distances are "pulled up" or exaggerated on the order of two to five times the horizontal, but this allows landforms to pop out all the more readily. Another distortion—in this case

*Two sources for an inexpensive stereoscope with plastic lenses are: Nasco, 901 Janesville Avenue, Fort Atkinson, Wisconsin 53538 and Ward's Natural Science Establishment, Inc., P. O. Box 1712, Rochester, New York 14603.

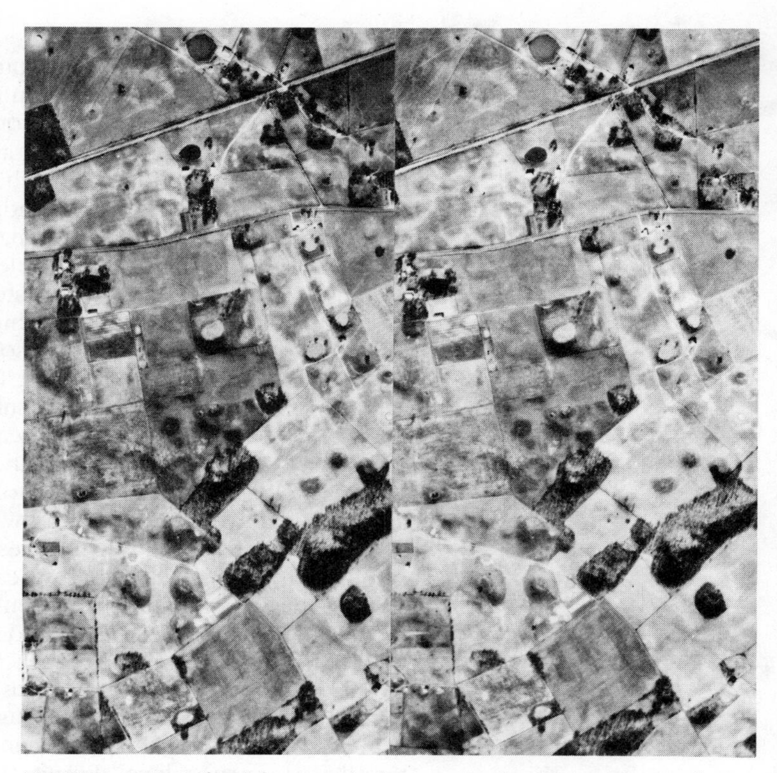

FIGURE 1–3. Vertical aerial stereopair photographs of sinkholes (Chapter 6) in cherty limestone near Oakland, southwestern Kentucky. Lacking a stereoscope, stare straight down, with each eye fixed on a common point on each photograph. Command your brain to bring the two points together as one for a 3-D effect. The darker sinkholes are wooded. The light nonwooded sinkhole near the center of each photograph is about 500 feet (152 m) wide. Figure 6–2 is a topographic map of a similar area whose center is about 8 miles (13 km) to the right or east-northeast. (U. S. Geological Survey photographs Kentucky 2 A–B).

horizontal—occurs along the margins of a photograph because of the inadequacy of any photographic lens near its margins.

If you're concerned about scale on an aerial photograph, check a topographic map of the same area. Simply compare two known points on each, and you can work out a ratio between the two.

Again, as with topographic maps, to sharpen your landform identification eye you might page through compilations of aerial photographs that explain the landforms you see on them. Mollard's *Landforms and Surface Materials of Canada,* with several hundred photographs, is particularly comprehensive. Others are Wanless' *Aerial Stereo Photographs* and the *Atlas of Landforms* by Curran and others. The *Atlas* combines vertical and oblique aerial and ground photographs with topographic maps. Shelton's *Geology Illustrated* provides numerous, high-quality oblique aerial as well as ground photographs of landforms.

Access to comprehensive collections of aerial photographs may not be as ready as for topographic maps. You might try county or local offices of such federal agencies as the Soil Conservation Service, the United States Forest Service, and the Agricultural Stabilization Conservation Service. Aerial photographs can be expensive, but, should you wish, they can be obtained readily from one of several federal agencies. Write for a Geographic Search for Aircraft Data Inquiry Form from the National Cartographic Information Center (see page 6). After returning the completed form, you will be sent a price list and order form to purchase your desired photographs.

I'd like to recommend an approach to the effective learning and appreciation of landforms. Before a trip, skim Chapters 2 to 9 and look over the keys in Chapter 10. If you have access to Photo-Geographic International's *Photo-Atlas of the United States* and Snead's *World Atlas of Geomorphic Features,* check them out. The *Photo-Atlas* shows complete photographic coverage of the United States by Landsat satellite photographs. The scale is too small to see many landforms, but you will acquire a useful perspective. Snead's *World Atlas* provides major locations of landforms on a worldwide basis, as the title implies. And, if you have time, look over some topographic maps and aircraft photographs as I have already suggested.

Selected Readings

CURRAN, H. A., P. S. JUSTUS, E. L. PERDEW, and M. B. PROTHERO. *Atlas of Landforms* (2nd ed.). New York: John Wiley and Sons, Inc., 1974.

DEBRUIN, RICHARD. *100 Topographic Maps.* Northbrook, Illinois: Hubbard Press, 1970.

HUNT, C. B. *Natural Regions of the United States and Canada.* San Francisco: W. H. Freeman and Company Publishers, 1974.

MOLLARD, J. D. *Landforms and Surface Materials of Canada: A Stereoscopic Airphoto Atlas and Glossary* (4th ed.). Regina: Commercial Printers, Ltd., 1975.

PHOTO-GEOGRAPHIC INTERNATIONAL. *Photo-Atlas of the United States.* Pasadena, California: Ward Ritchie Press, 1975.

SHELTON, J. S. *Geology Illustrated.* San Francisco: W. H. Freeman Company Publishers, 1966.

SHIMER, J. A. *Field Guide to Landforms in the United States.* New York: Macmillan, Inc., 1972.

SNEAD, R. E. *World Atlas of Geomorphic Features.* New York: Robert E. Krieger Publishing Co., Inc. and Van Nostrand Reinhold Co., 1980.

UPTON, W. B., JR. *Landforms and Topographic Maps.* New York: John Wiley and Sons, Inc., 1970.

WANLESS, H. R. *Aerial Stereo Photographs* (2nd ed.). Northbrook, Illinois: Hubbard Press, 1969.

WYCKOFF, JEROME. *Rock, Time, and Landforms.* New York: Harper and Row, Publishers, Inc., 1966.

2 STREAM-RELATED LANDFORMS

Running water, aided by the downslope movement of soil, sediment, and rock under the force of gravity, sculpts most of the landscape. Streams work geologically in two ways: They cut channels, and they serve as conveyor belts in removing mineral and rock debris brought to them by downslope movement. Channels are cut by the abrasive action of sand and gravel (see Chapter 16 for definitions of these and related sediment terms) grinding against the bottom and sides of a stream channel and by the force of the running water itself. As downcutting oversteepens valley walls, valleys widen—slopes retreat away from stream channels—chiefly by downslope movement, that is, landsliding and related processes (see Chapter 7). Imagine a stream valley where only channel downcutting—not downslope movement—operates. Result: a vertically walled gorge or canyon. These form only where resistant rocks do not readily succumb to gravity. *Sheet erosion,* resulting from the flow of rainwater in broad sheets and not in well-defined channels, assists in the movement of rock material down valley slopes.

Streams deposit sediment as well as erode, and a stream's course can be divided into three segments relative to erosion and deposition: an upper or upstream reach, where erosion prevails; a middle reach, where erosion and deposition are somewhat equal; and a lower or downstream reach, where deposition prevails.

Narrow Valleys

Narrow valleys (Figure 2–1) are narrow-bottomed—streams fill their bottoms—and V-shaped in cross-valley profile, as viewed up- or down-valley. Not necessarily narrow as the name implies, they may be half a mile (0.8 km) or more wide at their tops. The Grand Canyons of the Colorado or Yellowstone Rivers are "narrow" valleys. Narrow valleys occur mostly where streams are actively downcutting, especially in the upper reaches. But they exist elsewhere, as where underlain by rocks resistant to stream erosion.

FIGURE 2–1. Narrow valley, showing characteristic V-shaped cross-profile and waterfall. Grand Canyon of the Yellowstone River from Artist's Point, Yellowstone National Park, northwestern Wyoming.

Waterfalls, cascades, and rapids characterize narrow valleys where the *gradient* or slope of the stream bed steepens abruptly and stream velocity accelerates. At **waterfalls,** streams plunge vertically or nearly so. Niagara Falls is created where the Niagara River drops from a resistant dolostone into a plunge pool carved from relatively weak shales and limestones underlying the dolostone. **Cascades** are a series of small waterfalls or very steep rapids. **Rapids** are simply areas of faster-than-normal flow; they may develop from waterfalls. Waterfalls, cascades, and rapids commonly migrate upstream as undercutting of weak rocks causes overlying resistant rocks to collapse.

Potholes are holes—usually deeper than wide—in stream beds of rock at waterfalls, cascades, and rapids. They are ground out by the abrasive tools of sand, pebbles, cobbles, and boulders spinning and whirling about in the racing, turbulent water.

Broad Valleys

Broad valleys are broad and flat-bottomed—much wider than they are deep. They occur where side-to-side or lateral cutting prevails over downcutting, especially in the lower reaches of streams. Much of the Mississippi River flows through a broad valley.

FIGURE 2–2. Oblique aerial photograph of a broad valley with a meandering stream, point bars, meander scars, and flood plain. Laramie River, southeastern Wyoming. (Photograph 7 by J. R. Balsley, U. S. Geological Survey.)

One of nature's grandest designs—a broad stream valley complex—is best seen from the air (Figure 2–2). Dominating the design is the sinuous, meandering stream channel. Straight channels are not the norm, and in broad valleys they are rare. **Meanders**—wide loops or bends of the stream channel—develop in broad valleys because the gradient is low and the terrain easily erodible. Higher stream velocity on the outside of a meander causes erosion there by undercutting and collapse of cutbanks; lower velocity on the inside causes deposition there of stream sediment in arclike ridges or **point bars** at the point of the meander. These ridges are frequently emphasized by stream-flanked vegetation. (To test the high versus low velocity on the outside and inside of a meander, paddle a canoe through a meandering stream. Better still, try snorkeling with a dependable PFD [Personal Flotation Device] and you will experience the velocity differences even more intimately!) Combined lateral cutting and deposition allow meanders to shift back and forth across a valley, as well as to migrate—in time—down the valley because of the slope of the valley surface.

Now, what about those stream loops isolated from the main stream channel? If the downstream limb of a meander is held up in its migration down the valley by, say, resistant rock material, the upstream limb overtakes it and the meander is cut off from the main channel (Figure 2-3). As the ends of the abandoned meander are plugged with sediment, an **oxbow lake** is formed. This becomes a **meander scar** if the lake dries up or fills with sediment and vegetation. (Incidentally, the "billabong" of the unofficial Australian national anthem, "Waltzing Matilda," is an oxbow lake.)

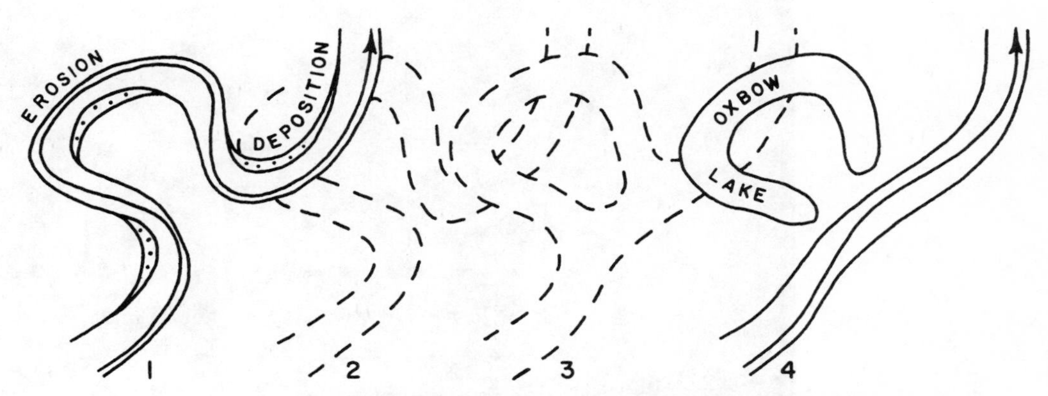

FIGURE 2-3. Stages in the cutting off of a meander and formation of an oxbow lake. A stream erodes on the outside of meanders and deposits sediment on the inside to form point bars (stage 1, areas stippled). The neck of the meander gradually narrows and severs as the ends of the abandoned meander are plugged with sediment. An oxbow lake is created. The sinuous line within the channel in stages 1 and 4 traces the path of highest stream velocity.

The relatively flat part of a broad valley, inundated during floods, is the **flood plain** (Figure 2-2). It is generally fertile for agriculture because increased amounts of organic matter, as well as finer sediment, settle on its surface as succeeding flood waters recede. Frequently, it stands out in photographs because of the checkerboardlike pattern of fields (see Figure 1-2). During floods, low ridges of silt and sand flanking a stream channel—**natural levees**—are deposited as stream velocity is checked abruptly near the channel. With each flood, the natural levees—and stream bed—are gradually raised above the **backswamp** or lower ground surfaced with silt and clay away from the river. From the air, natural levees are evidenced by tributary streams flowing parallel to the main stream for some distance before reaching breaches in the levees. Obviously, the flood plain develops from suspended sediment settling out from flood water. But it develops, too, as stream meanders shift laterally across a valley. So a flood plain results

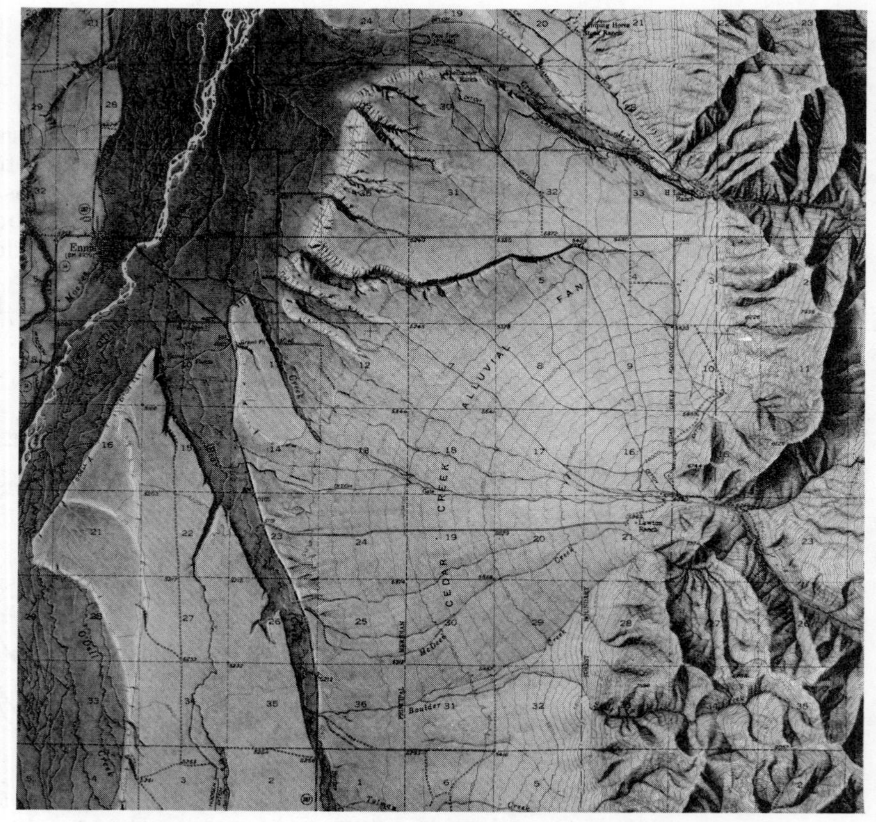

FIGURE 2–4. Shaded relief topographic map showing stream features, Ennis area, southwestern Montana. The Madison River is braided in the upper left, stream terraces are visible on the left, and the Cedar Creek alluvial fan dominates the center. The numbered squares of the grid are one mile (1.6 km) on a side. (From the U. S. Geological Survey Ennis Quadrangle, 1949.)

from erosion and reworking of stream sediment as well as from *overbank* deposition during floods.

On some flood plains of broad valleys, a stream consists of many interconnected channels—resembling the strands of a braid—instead of a single, meandering channel. Bars and islands abound. Such a **braided stream** (Figure 2–4) occurs where the sediment supplied to it is more than it can carry.

When a meandering, broad-valley stream undergoes a period of renewed downcutting or *rejuvenation* for any of the reasons discussed in Chapter 19, stream terraces or incised meanders may result. **Stream terraces** (Figures 2–4, 9–15), remnants of flood plains flanking valley walls and cut by lateral erosion as well as by downcutting, may be paired on either side of a valley or not. If not, resistant rock underlying a terrace may preserve it on one side of the valley, but the lack of such rock on the other side of

FIGURE 2-5. Incised meander.

the valley allows the stream to remove the other member of the pair. Stream terraces, which also may be cut into resistant rock, may form a complex at several levels, depending on the rejuvenating behavior of a stream or region. **Incised** (or **entrenched**) **meanders** (Figures 2-5, 9-8) are those incised or cut deeply into sediment or rock without associated terraces. Downcutting is the main erosive process here, with little or no lateral erosion occurring.

When a stream carrying considerable sediment enters standing water, its velocity is checked, the sediment is dropped, and a **delta** is formed. Ideally, this body of sediment is fan-shaped or triangular like the Greek letter delta from which the name is derived. The Nile delta in Egypt, whose profile is partly molded by waves and currents, fits the ideal delta shape. But few deltas do. The Mississippi delta, where stream processes dominate over those of shorelines, is highly asymmetrical with irregular margins. The fanning or spreading out of a delta results from the main stream splitting into several smaller streams or **distributaries** as the velocity slackens and sediment is deposited. New distributary systems or lobes generally are built during floods. A geologic challenge is recognizing an ancient delta, whose form may be seen rising subtly above a former lake bottom (see Figure 3-10, bottom).

Similar to deltas are **alluvial fans** (Figures 2-4, 2-6). Fan-shaped bodies of sediment at the base of steep slopes, they form where streams drop sediment as their velocities are checked by abruptly lessened gradients. Alluvial (*alluvium* is stream sediment) fans occur mostly in arid or semiarid regions where intermittent streams, flowing through mountain canyons, spill out onto a low-

FIGURE 2–6. Alluvial fans at the base of a mountain front, Mojave Desert, California. Compare with the alluvial fan in Figure 2–4. (Sketched from a U. S. Geological Survey photograph by J. R. Balsley.)

land. They fan out or spread as the main stream shifts laterally and divides into smaller distributaries. Alluvial fans, in which are deposited largely coarser sand and gravel rather than the finer sand and mud in deltas, are associated with narrow rather than broad valleys.

Landscape Change by Stream Erosion

So far, we have examined narrow and broad valleys as if they might be distinct entities. But each stream valley, to be sure, is always part of a larger drainage system that molds a particular landscape (Figure 2–7). A drainage system of a stream and its tributaries extends upslope toward a drainage divide—the boundary between adjacent drainage systems—by *headward erosion.* Each tributary valley lengthens its upper, or headwater, reach as sheet erosion converts to stream erosion at the head of the valley.

Geologists have conceptualized that stream erosion, accompanied by downslope movement of rock material and sheet erosion, causes successive changes in a landscape. No one has witnessed a complete series of stream-induced landscape changes at any one place, but different stages in the series of changes can be observed in different places. **Base level,** the lowest level to which a stream can erode, such as sea level—or, temporarily, a lake, reservoir, or resistant rock layer—is the controlling factor. The lower the landscape, or the closer it is to base level, the more it has changed. You should remember that interruptions in the progression of landscape changes may occur at any time.

In a humid region (Figure 2–7), the landscape passes from

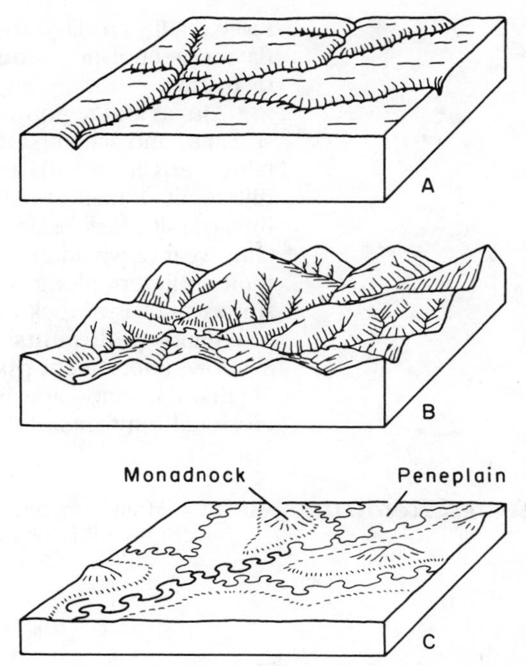

FIGURE 2–7. Landscape change in a humid region. *A*, early stage: Divides or boundary areas between streams are broad, tributaries are few, and relief is low; stream valleys are narrow, and streams have steep gradients with rapids and waterfalls (Figure 2–1). *B*, intermediate stage: Divides between streams are narrow, tributaries are numerous, and relief is highest; stream valleys are broad, and meanders and flood plains develop. *C*, late stage: The region is reduced to a low-relief surface (peneplain) near sea level with scattered erosional remnants (monadnocks); stream valleys are very broad, and flood plains are wider than stream meander belts.

one of low relief to high relief and back to low relief. Highest relief is associated with the stage of greatest number of tributaries. Divides between streams change from broad to narrow to much reduced. Streams initially have high gradients and flow through narrow valleys with their characteristic features. Later, gradients are reduced, and broad valleys with their characteristic features are produced. Finally, if no interruptions occur, the changing landscape becomes a low-relief erosional surface or **peneplain** with scattered erosional remnants or **monadnocks.** Flood plains are wider than the meander belts—defined by imaginary lines drawn tangent to the edges of the meanders—of the highly meandering streams on the low-relief surface.

In an arid region with through-flowing streams, landscape changes are similar to those in a humid region. However, landforms are more angular, the retreat of slopes is more evident, and wind action may be important. As the region is dissected, flat-lying resistant rocks leave wide tablelands or **plateaus** standing above more readily erodible areas. Plateaus reduce to smaller, steep-sided, flat-topped uplands—**mesas,** which reduce further to isolated hills or **buttes.**

In an arid region with internal drainage (no through-flowing streams) and accompanied by the formation of fault-block mountains (see Chapter 10) and intervening basins, the landscape alters differently. Initially, the linear fault-block mountains project above linear fault-block basins. Relief gradually decreases as the mountains wear down and the basins fill with sediment. Streams deposit their sediment along mountain fronts in alluvial fans. Erosional surfaces cut into rock along mountain fronts and veneered with sediment—**pediments**—widen as the mountain ranges diminish. Shallow, intermittent **playa lakes** in the basins accumulate salts. Finally, the landscape is reduced to a low plain with scattered erosional remnants of the once lofty mountain ranges.

Selected Reading

MORISAWA, MARIE. *Streams: Their Dynamics and Morphology.* New York: McGraw-Hill Book Company, 1968.

3 GLACIER-RELATED LANDFORMS

You are flying in a high-altitude jetliner over southern Alaska, smirking slyly at the copy of *Mad* magazine glued to your lap. The captain's voice crackles through the aircraft: "Good afternoon, ladies and gentlemen. This is Captain Barnes. We are on schedule, and presently crossing the Alaska Range. Thought you might want to look at the glaciers that are directly beneath us."

"Look, there they *are!*" exclaims an elderly gentleman as he points excitedly toward the window.

The *Mad* magazine spills to the floor, and you eagerly press your nose against your window. There, like striped, snaky locks of Medusa's head, you see them—issuing in all directions from the lofty snow fields. Fantastic! . . .

Now, if I have your attention, let's take a closer look at glaciers and the landforms they shape.

Glaciers and Glacial Ice

Glaciers are thick masses of ice, moving now or having moved in the past. Icebergs don't qualify, although they may calve from a glacier that enters the sea or a lake.

Two main types of glaciers exist: valley and continental. **Valley** or **alpine glaciers** (Figure 3–1) are confined to mountain valleys and are, literally, streams of ice flowing down these valleys. At high altitudes in many parts of the world—even in the tropics—they are several hundred feet to many tens of miles long and a few hundred to a few thousand feet thick. Where valley glaciers spill out onto low-lying terrain at the feet of mountain fronts and form broad lobes, they are called **piedmont glaciers.**

Continental glaciers or *ice sheets*—small ones are *ice caps*—cover large parts of continents. Antarctica and Greenland are blanketed by them; here, glaciers are 10,000 feet (3000 m) thick or more, enough to bury entire mountain ranges. Can you imagine ice *two miles* (3.2 km) or more in thickness?

FIGURE 3–1. Valley glaciers, latitude 66°42′ N, longitude 65°47′ W, southeastern Baffin Island, Canada. Visible are crevasses; lateral, medial, recessional, and end moraines; outwash deposits; glaciated valleys; cirques; and arêtes. (Aerial photograph T332R-150 © 1949 Her Majesty the Queen in Right of Canada, reproduced with permission from the National Air Photo Library, Department of Energy, Mines and Resources, Canada.)

Today, glaciers cover about 10 percent of the earth's land surface; 90 percent of that ice is on Antarctica. But during the Pleistocene Epoch (see Chapter 11), beginning about 2 million years ago, glaciers were more extensive. Their margins, however, fluctuated several times, expanding and encroaching at one time, melting back at another. Then, 20,000 to 15,000 years ago, glaciers oozed out to cover the greatest acreage—about 30 percent of the earth's land surface. Ice covered most of northern North America—including the northern United States and Greenland—northern Europe and Asia, Antarctica, and southernmost South America. Glaciers on Antarctica and Greenland today represent the most conspicuous remnants of the great Pleistocene ice sheets.

What is needed for glacial ice to form? Simply, a cool, humid climate, and more snow accumulating than melts each year. Glacial ice is made from considerable snow, by pressure—compaction—

and recrystallization. It is made, in fact, by a metamorphic process (see Chapter 17). Highly porous, low-density masses of snow are pressed together. Points of snow crystals melt from the pressure, and water freezes and recrystallizes in the intervening spaces which are subject to lesser pressure. In time, the snowflakes transform into tiny grains. (Such granular snow is what you find in old snowbanks in late winter or earliest spring.) Further pressure and recrystallization produce dense, nonporous, glacial ice of interlocking ice crystals—basically, a metamorphic rock. This transformation, from snow to glacial ice, may take from a year or so to tens of years, depending on the climate.

That glaciers move is particularly obvious in the flow lines seen on the surface of valley glaciers and emphasized by rock debris (see Figure 3-1)—the striping of Medusa's locks mentioned at the beginning of the chapter. But how can ice, that brittle substance in your glass, "flow?" Beneath an upper zone about 100 to 200 feet (30 to 60 m) thick where **crevasses**—cracks or fissures—form as the glacier plunges over steep spots in the valley floor, the ice is a plastic, yielding substance—like a thick, heavy fluid, if you will. Folds in the ice seen near the fronts of glaciers attest to this. As a heavy mass of ice succumbs to gravity, flowage occurs by gliding along tiny planes within ice crystals as well as by ice crystals rotating and sliding past one another. Glaciers move, too, by slippage along their margins, shown by grooved, scratched, and polished rock exposed after the ice has melted away. At glacier fronts, thinner, brittle ice moves by sliding along shear planes.

Glacial ice moves about as fast as ground water (see Chapter 6) percolating through porous rock or sediment. Most glaciers "race" along at rates of less than an inch to several feet per day; surging glaciers may occasionally spurt along at rates of a few to several hundred feet per day.

Work of Glaciers

Glaciers wear down rocks and sediment mostly by *plucking* and *abrasion.* Melt water seeps into cracks in rock beneath and along the sides of a glacier, freezes, and expands. Blocks of rocks are wedged loose, frozen to the base of a glacier, and *plucked* out as the glacier moves along. The plucked rocks, like grains in sandpaper, *abrade*—rasp, grind, groove, scratch, polish—the rocks on which the glacier passes as well as themselves. Many gravel-sized particles have characteristic grooved, ground faces or facets, crudely similar to those on cut gemstones (Figure 3–2), and differ from particles found in streams, which have curved surfaces.

Glaciers carry rock material largely in suspension at the base, sides, and top of valley glaciers and mostly near the base of continental glaciers. Some rock material is also bulldozed in front.

When glaciers melt, rock material is deposited. Material let down directly by the melting ice is **till** (Figure 3–3), a poorly sorted mixture—lacking uniformity of particle size—of particles ranging from boulders through clay. Till lacks distinct layering or

FIGURE 3–2. Glacially grooved and faceted dolostone cobble, from near Inkster, northeastern North Dakota. Cobble is 2.9 inches (7.4 cm) high.

FIGURE 3–3. Glacial till, exposed in a stream cutbank, showing general lack of bedding. A boulder protrudes from the till at the left of the photograph. The exposure is about 35 feet (11 m) high.

FIGURE 3–4. Glacial meltwater sand and gravel or outwash sediment, showing distinct cross-bedding. The pebble in the lower right is 1.8 inches (4.6 cm) wide.

bedding and contains the sculpted tools of abrasion—faceted, scratched, grooved gravel particles. Sediment worked and transported by melt water is better sorted and displays good bedding: the sand and gravel of melt water streams (Figure 3–4) and the silt and clay of glacial lakes.

Landforms of Valley Glaciers

Valley glaciers occupy former stream valleys and, in so doing, deepen, widen, and somewhat straighten them, or at least smooth out many irregularities (Figure 3–5). A cross-valley profile of a **glacial valley** or trough is characteristically U-shaped (Figure 3–6) unless later sediment fill has flattened the valley floor. A profile along its length is likely to show steplike irregularities with scooped-out **rock basins** commonly containing lakes. Along coasts, flooded glacial valleys form **fiords.** Tributary glaciers don't scour as deeply as main glaciers, so when ice melts they form higher **hanging valleys** from which waterfalls often plunge.

At the head or upper end of a glacial valley is a half-bowl–like depression or **cirque** (Figures 3–5, 3–7; see also Figure 3–1) where a glacier originates. Its cliffed walls form and extend upvalley—headward—by ice wedging and plucking, the washing

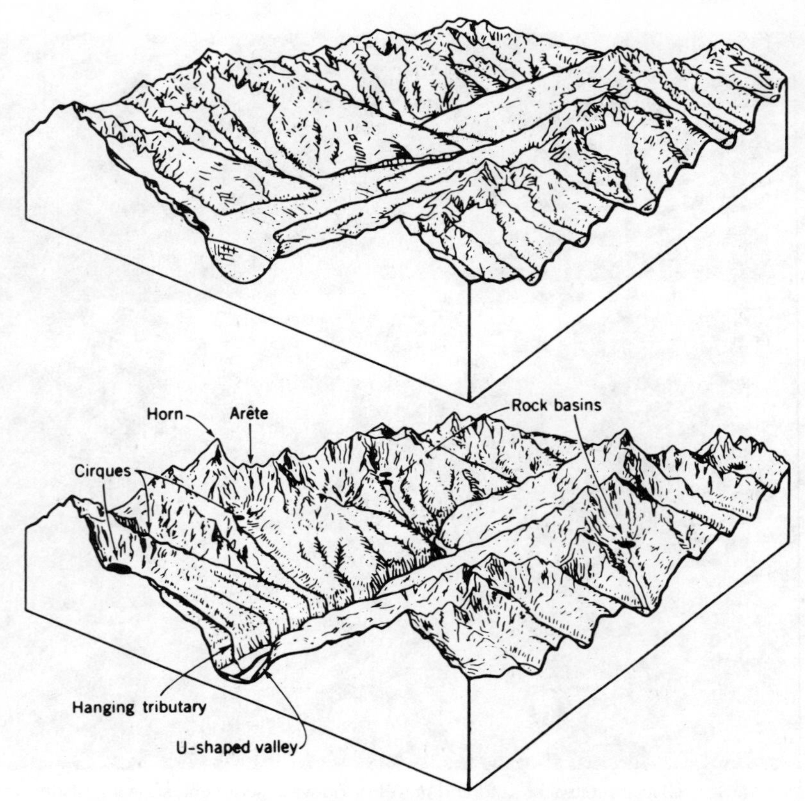

FIGURE 3–5. Erosional features of valley glaciers. As valley glaciers occupy former stream valleys (*top*), they deepen, widen, and straighten them somewhat. After glaciers melt, characteristic landforms (*bottom*) evidence their former existence. (From R. F. Flint and B. J. Skinner, *Physical Geology,* 1974, Figure 11.12, used with permission of John Wiley and Sons, Inc.)

action of melt water, avalanching, and rockfalling. Abrasion scours a rock basin at its floor, which is frequently occupied by a lake. Two cirques coming together by headward erosion on opposite sides of a divide produce an **arête** or sharp-crested ridge; three or more create a pyramidal peak or **horn,** of which the Swiss Matterhorn is a famous example.

Several landforms arise from the deposition of rock debris. Ridges along the sides of valley glaciers are **lateral moraines** (Figures 3–1, 3–8), formed as debris accumulates by valley wall erosion, avalanches, and rock falls. Where a tributary glacier merges with the main glacier, lateral moraines merge as well to form **medial moraines** (see Figure 3–1), several of which may ribbon a glacier's surface. Medial moraines rarely remain in glacial valleys because of stream reworking during and after glacier melting. Between lateral moraines on the valley floor is an undulating surface of low hills and closed depressions—a blanketlike **ground**

FIGURE 3–6. Glacial valley with a characteristic U-shaped cross-valley profile. The bottom of the valley is somewhat flattened because of later sediment fill. Yosemite Valley, Yosemite National Park, eastern California. El Capitan is the cliff on the left, Bridalveil Fall drops from a hanging valley on the right, and Half Dome—an exfoliation dome (Chapter 23)—is in the center far distance. (Photograph 748 by F. E. Matthes, U. S. Geological Survey, 1923.)

FIGURE 3–7. Shaded relief topographic map of terrain sculptured by valley glaciers, showing glaciated valleys, cirques, rock basin lakes, and an end moraine fringing Turquoise Lake (lower right). The numbered squares of the grid on the right side of the map are one mile (1.6 km) on a side. Near Leadville, west-central Colorado. (From the U. S. Geological Survey Holy Cross Quadrangle, 1949.)

FIGURE 3–8. Field sketch of Athabasca Glacier, Jasper National Park, southwestern Alberta, Canada; June 15, 1959. Glacier-fringing lateral moraines are conspicuous.

moraine of till let down by the melting ice. Lateral moraines often merge with an **end** or **terminal moraine,** an arclike ridge of till that forms at the lower end or terminus of a glacier. It forms when an ice front stabilizes for a time: the glacier wastes away at the same rate as it is nourished. An end moraine may dam up melt water, creating a lake on its upvalley side (see Figure 3–7). If the ice front recedes and restabilizes, a **recessional moraine** (see Figure 3–1)—upvalley from the end moraine—is laid down. Away from and below the end moraine is a relatively flat **outwash plain** of stream sediment "washed out" from the glacier by melt and rain water. A body of outwash confined to a valley is a *valley train.*

Streams flowing on the glacier or in tunnels within or beneath it lay down sand, gravel, and silt within their twisting, icy channels. When the ice melts, the channel sediment is let down to form sinuous ridges—**eskers**—seemingly loping over the uneven glaciated terrain. Eskers pass downstream into deltas—raised above the surrounding terrain after melt water lakes disappear—and outwash plains. If a glacial stream dumps its sediment into a restricted opening in the ice (a crevasse), sediment let down upon melting assumes the configuration of a steep-sided hill—**kame**—or a relatively straight ridge—**crevasse filling.**

As glaciers sculpt a mountainous region, its topography takes on a distinctive appearance. Rugged, *angular* landforms predominate. Characteristic U-shaped valleys are interspersed with sharp-crested divides and jagged peaks (see Figure 3–5).

Landforms of Continental Glaciers

Hardly spectacular by most standards, the landforms of continental glaciers are generally inconspicuous and are often missed by the casual traveler. But they do cover the acreage. Such landforms in northern North America, for example, can be found over an area greater than the size of the Antarctic ice sheet, attesting to the extensive ice coverage. What commonly snares the eye—perhaps because of the monotony—is an undulating or hilly terrain with closed depressions, many filled with ponds, marshes, bogs, or lakes. Stream drainage systems, on the younger surfaces or where till is thick, are poorly developed or nonexistent. The terrain is the kind you would see in northern Minnesota and Wisconsin, most of Michigan, and western New York. On the surface of the terrain and embedded in it are numerous boulders, some of which are house-sized, differing in composition from the rock near which they are found. Examples would be boulders of granite, gneiss, and dolostone resting on or near sandstone and shale. Obviously transported some distance from their source, these **erratics** were one of the first proofs for glaciers. What else could move huge boulders, especially those of house-sized proportions, great distances?

Erosional landforms of continental glaciers are barely noticeable in many places. They consist of large, nearly flat low-lands of grooved and scratched rock stripped of most or all of its soil and sediment cover, as in eastern Canada and northern Europe. In some places there are asymmetrical rock knobs or **sheep rocks** (Figure 3–9)—the French equivalent is *rôche moutonnée*—formed by abrasion on the less steep, upglacier side and by plucking on the steeper, downglacier side. (Downglacier is the side or direction toward which the glacier was moving.) Such rocks, then, are useful in reconstructing past ice movement. More conspicuous landforms are the rock basins scooped out of valleys by separate lobes along the scalloped margins of ice sheets. Occupying such basins are the Great Lakes and the smaller, long and narrow *finger lakes,* as the north-trending Finger Lakes of west-central New York.

Moraines of continental glaciers (Figure 3–10) are similar to those of valley glaciers, but lateral and medial moraines are lacking. End moraines (Figure 3–11) may be multiple and complex, forming, in places, a few- to several-miles-wide band of concentric ridges

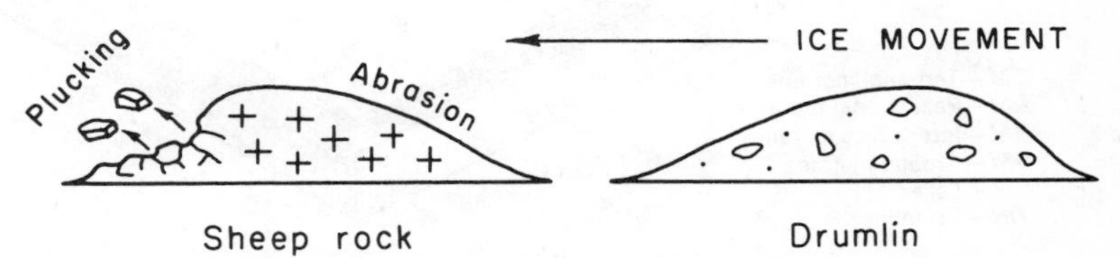

FIGURE 3–9. Generalized profiles of a sheep rock and drumlin in relation to the direction of ice movement. Sheep rocks are knobs of hard rock whereas drumlins consist mostly of till.

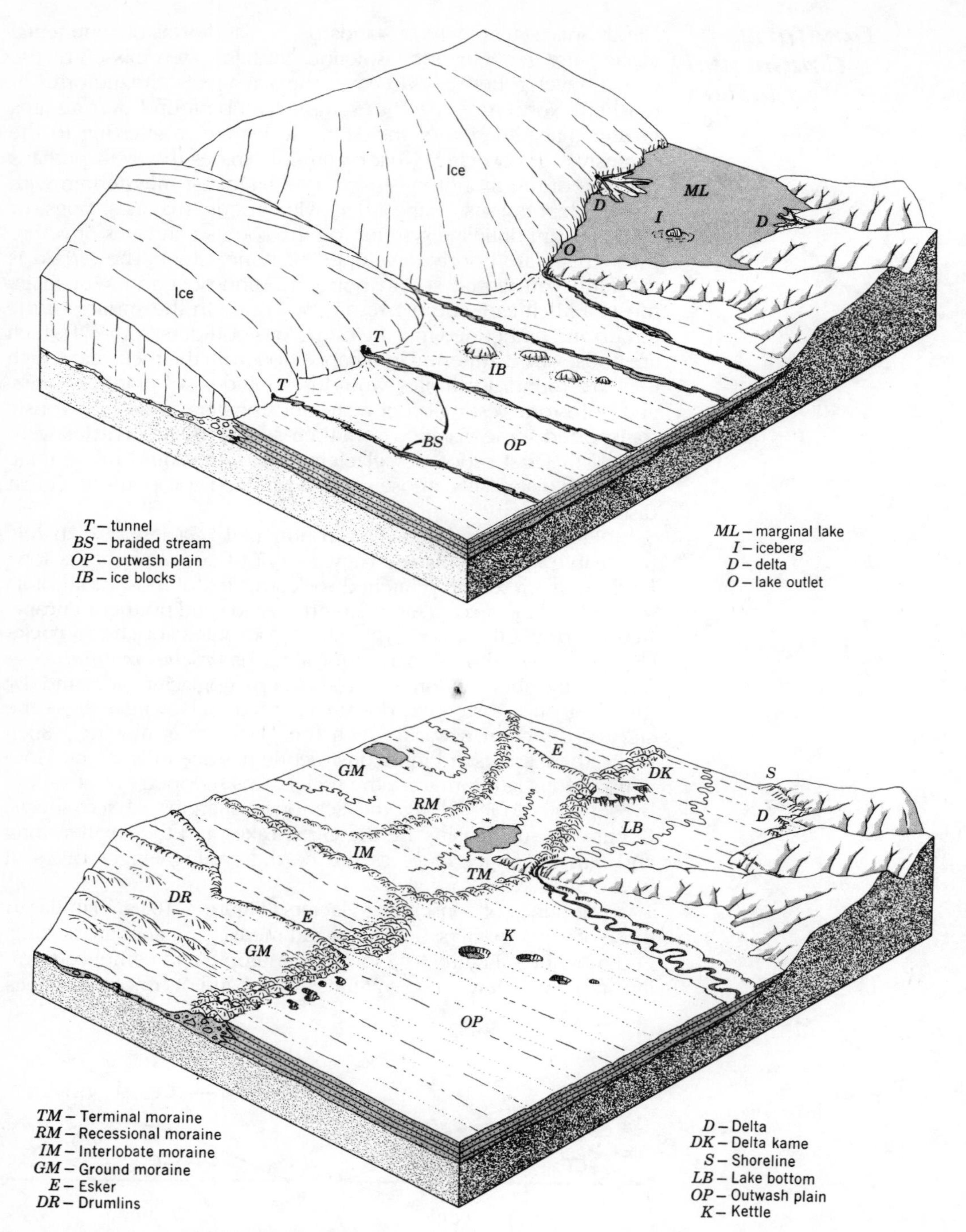

FIGURE 3-10. Depositional features of continental glaciers. Glacier melt water forms features at the margin of the ice (*top*). Upon complete melting, other depositional landforms (*bottom*) are exposed. (From A. N. Strahler, 1975, *Physical Geography,* 4th ed., Fig. 31.3, with permission; copyright © 1975 by John Wiley & Sons, Inc.)

FIGURE 3–11. End moraine, with many glacial erratics on the surface.

FIGURE 3–12. Ground moraine with glacial erratics, showing less relief than that of end moraine (Figure 3–11) and a water-filled depression or swale.

that produces a washboard effect. Continuations of end moraines between adjacent ice lobes are *interlobate moraines.* Ground moraine (Figure 3–12) is the most extensive type, and has less relief than do end moraines. Its more subdued relief derives from uneven laying down of till or from an uneven surface that underlies it. On some ground moraine occur streamlined, asymmetrical hills—**drumlins** (Figures 3–9, 3–13)—that resemble inverted teaspoons (minus the handles) in plan view or are more elongate. Most are of till, others contain sand and gravel, and some are rock-cored with a veneer of till. Molded by moving ice and oriented parallel to its movement, they appear at right angles to

FIGURE 3-13. Surface overridden by a continental glacier, showing a drumlin field and numerous lake-filled depressions; latitude 62°45′ N, longitude 106°00′ W, east of Great Slave Lake, District of Mackenzie, northwestern Canada. Several drumlins show blunt ends pointing toward the horizon indicating the direction from which came the glacier that formed them (Figure 3-9). (Aerial photograph T82R-160 © 1946 Her Majesty the Queen in Right of Canada, reproduced with permission from the National Air Photo Library, Department of Energy, Mines, and Resources, Canada.)

end moraines. Drumlins are usually less than 0.5 mile (0.8 km) long and 150 feet (46 m) high. The more gentle slope along a drumlin's length points in the direction the ice was moving. Drumlins generally occur in fields or swarms, as in southeastern Wisconsin and in west-central New York north of the Finger Lakes.

Other features generally formed by direct contact with stagnant, nonmoving ice—eskers, kames (Figures 3-14, 3-15), crevasse fillings, and outwash plains—are similar to those of valley glaciers. Eskers, however, may be longer and the outwash plains more extensive. Blocks of ice, separated from the main mass of ice, may become partly or wholly buried in outwash sediment or the till of ground moraine. Upon melting, steeper-walled depressions—**kettles** (Figure 3-16)—form that may become lakes, swamps, or bogs.

As a glacier melts, considerable water accumulates as melt water streams that form most of the features we've just considered. Other melt water streams simply siphon off the water elsewhere,

FIGURE 3–14. Dahlen esker, near Dahlen, northeastern North Dakota. (Photograph by J. R. Reid.)

FIGURE 3–15. Woods-covered kame.

FIGURE 3–16. Forest-flanked kettle.

FIGURE 3–17. Aerial view of the Glacial Lake Agassiz plain, near Grand Forks, northeastern North Dakota. (Photograph by J. R. Reid.)

FIGURE 3–18. Ground view of the fertile Glacial Lake Agassiz plain, showing a field of sunflowers. Near Fargo, southeastern North Dakota.

and when flow ceases, abandoned **melt water channels** remain behind with perhaps *underfit streams*—those too small for the valleys they occupy. Much of the melt water accumulates in lakes, as already mentioned.

With the advance of glacial ice, stream courses are diverted or blocked altogether, forming marginal streams or lakes. The present courses of the Missouri and Ohio Rivers approximate the southernmost extent of continental glaciers and document such diversion. A major drainage blockage was that of the Red River of the North. This created Lake Agassiz, largest of the marginal lakes in North America, named after a famous nineteenth-century zoologist and glaciologist. Its extent, varying as the ice margin fluctuated, reached an area greater than that of all of the Great Lakes combined and occupied parts of Manitoba, Ontario, and Saskatchewan as well as eastern North Dakota and northwestern Minnesota. Lakes Winnipeg, Winnipegosis, and Manitoba represent remnants of this once great lake. Today, the lake is evidenced by a vast, highly fertile, flat **lake plain** (Figures 3-17, 3-18) fringed by several ancient shorelines—pitted with sand-gravel pits in places—marking various positions of the lake's edge.

Selected Readings

BAILEY, R. H., and The Editors of Time-Life Books. *Planet Earth: Glacier.* Alexandria, Va.: Time-Life Books, Inc., 1982.

DYSON, J. L. *The World of Ice.* New York: Alfred A. Knopf, Inc., 1966.

POST, AUSTIN, and E. R. LACHAPELLE. *Glacier Ice.* Seattle: University of Washington Press, 1971.

PRICE, R. J. *Glacial and Fluvioglacial Landforms.* New York: Hagner Publishing Company, 1973.

SCHULTZ, GWEN. *Ice Age Lost.* Garden City, New York: Anchor Press, 1974.

4 SHORELINE-RELATED LANDFORMS

Shorelines, where land and water meet, are intriguing places to watch conflicting destructive and constructive geologic processes at work. Battering waves tear apart rocks, grind them down with their abrasive tools of gravel and sand, and, with longshore currents, carry the rock debris away. Earth movements may raise shorelines for renewed attack by wave erosion. Streams provide sediment for building the land out into deltas or for molding, by waves and longshore currents, into bars, barrier islands, and the like. So landforms along shorelines, as elsewhere, take shape by tearing down as well as by building up. But, as we will see at the end of the chapter, many landforms along shorelines were built by manipulators other than waves and currents.

Work of Waves and Currents

Waves, most important geologically in the ocean, are generated by wind. Wave motion transfers the wind energy from one place to another. Offshore in deep water, water particles move negligibly; it's the form of the waves that passes forward. In a field of waving grain, the stalks and heads bend slightly forward but return to their original positions—the stalks being firmly rooted in place—as a wave sweeps through the field.

A wave's energy depends on its length—the distance from one crest or high point to another—and on its height—the distance from a crest to a trough (the lowest point). Waves "feel," disturb, or erode the bottom to a depth equal to about one-half of the wave-length. So, waves with greater wave-lengths can erode to greater depths. As waves approach the shore—at a depth less than one-half of their wave-length—they slow down, crowd together, heighten, oversteepen, curl over, and break. *Breakers,* "broken waves," tend to form at a depth of one to one and one-half times the wave height. Nearshore ocean waves—except during severe storms—are seldom over 20 feet (6 m) high. So the **surf** or **breaker zone** of "broken water" (Figure 4-1) normally is

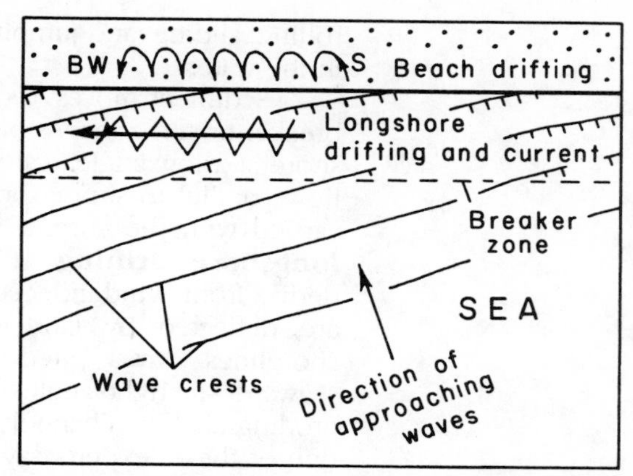

FIGURE 4–1. Beach drifting (curve pattern with arrows) and longshore drifting (zig-zag pattern with arrow) of sediment caused by waves approaching the beach (stippled) at an angle. Within the breaker zone (tick marks on wave crests) is the longshore current (straight line with arrow) that drives longshore drifting. Swash (*S*) and backwash (*BW*) drive beach drifting.

limited to a narrow band from the shoreline to a depth of 20 to 30 feet (9 m).

Where waves slam a rocky shore, the impact of the water smashes and loosens rocks, particularly if they are already fractured or cracked. Water driven into cracks compresses air that widens existing cracks and wedges and pries apart rock blocks. Spray-producing *blowholes* along sea cliffs are dramatic places where wave-induced compressed air is released. Storm waves move rock or concrete blocks weighing many hundreds of tons.

Abrasion—as in stream and glacial erosion—is the grinding of rock against rock, and a particularly effective wave erosion process. Sand and gravel particles—the tools of abrasion—grind against themselves as well as against rocky cliffs. Waves also erode, but insignificantly, by dissolving rock, especially limestone and dolostone.

Waves more powerfully erode jutting headlands or promontories than they do intervening bays. As a wave front approaches a headland, that part nearest the headland—where water is shallower—feels bottom first and is retarded. Those parts of the wave front on either side of the headland, however, proceed at normal speed, and are bent around until they parallel the shore. Wave bending concentrates energy and erosion on the headland and partly accounts for irregular shorelines becoming smoother and straighter with time.

As breaking waves splash a low shore at a low angle (less than at a right angle) the uprushing water or *swash* carries sediment up the beach slope and returns it directly back down as the *backwash,* pulled by gravity. In a continuous series of small, curved, swash-backwash paths, then, sediment moves along a beach by **beach drifting** (Figure 4–1). Coarser grains move by rolling, sliding, and jumping, and finer ones are held suspended in the water.

Sediment moves, too, in the deeper, turbulent water of the breaker or surf zone. As breaking waves once again approach the shore at a low angle, water piles up at first and then is forced to flow parallel to shore, forming a *longshore current.* Sediment is carried with the current in short, back-and-forth movements—**longshore drifting** (Figure 4–1). Longshore currents also derive from wind-induced oceanic currents whose movements are deflected by land masses and tend to nearly parallel shorelines. Water piled up from breaking waves breaks out seaward—at right angles to shore—as *rip currents,* capable of eroding shallow channels in the sea floor. If you are caught in one of these, experts say, swim parallel to shore to escape the narrow current. You can recognize them by gaps in breakers and streaks of darker, deeper water moving seaward, often with foam or floating objects.

Although most sediment is coarse and moves within the surf zone and beach, some finer sediment moves outside of the surf zone by the action of waves and currents. Oscillating waves move particles back and forth, but net movement is offshore as the particles are pulled down the sloping sea bottom by gravity. Each particle is sorted out from the rest according to the energy required to move it. When transporting wave action is too feeble, the particle comes to rest. Coarse particles form steeper slopes, finer particles more gentle slopes. If the available sediment reaches a balance with the prevailing wave and current energy, the shoreline profile—from shallow to deep water—tends to become concave upward, steeper near shore and more gently sloping in deeper water.

Other water movement is of less geologic significance than waves and longshore currents. Seismic sea waves, termed by the Japanese *tsunamis,* are set off by tremors from earthquakes, volcanic eruptions, or submarine landslides. Racing at speeds of a few to several hundred miles per hour, they can produce severely eroding surf—although lasting briefly and occurring rarely—that may destroy human life and property as well. Places like the Hawaiian Islands are particularly vulnerable to tsunamis. Tides are important geologically, mostly because they allow wave erosion to occur at more than one level. In narrow bays and inlets, tidal currents can loosen and move sediment.

Landforms of Marine Erosion

Breakers slamming directly against a shore promote landforms of erosion. Such landforms are fabricated more readily if the shore rocks are poorly solidified or well fractured. **Sea cliffs** (Figure 4–2) are formed not only by wave erosion but also by landsliding as steep, undercut slopes are oversteepened. **Hanging valleys,** analogous to those in areas of valley glaciation, abruptly terminate at sea cliffs where stream valleys meet the sea. Recesses in sea

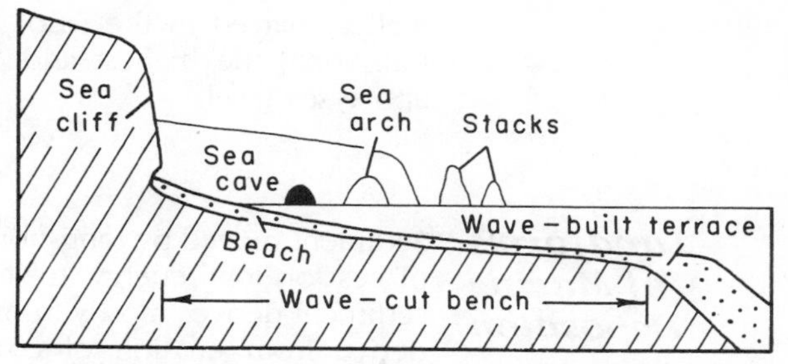

FIGURE 4–2. Profile of a shoreline of primarily erosion and associated features.

FIGURE 4–3. Shoreline of primarily erosion, showing wave-cut cliffs and many stacks. Mineral-rich black sand accumulates on the small beach. Curry County, southwestern Oregon. (Courtesy of Oregon State Highway Department, photograph 6674.)

cliffs may hollow out to **sea caves,** which, on opposite sides of a headland, may be excavated through to form a **sea arch. Stacks** (Figure 4–3), isolated pinnacles or small islands, result from the collapse of sea arches or the erosion of vertically fractured headlands.

As wave impact and abrasion pound, saw, and grind the sea cliff back, a **wave-cut bench** or platform develops at its base. Exposed at low tide in some places, the wave-cut bench widens until a critical width absorbs most of the wave energy. Thereupon, a beach forms along the low-energy shoreline, and the sea cliff diminishes as weathering and landsliding—and perhaps stream action—continue their effects. In regions of coastal uplift, wave-cut benches raised above sea level are called *marine terraces.* Several of these, raised during the late Pleistocene, are well displayed in the Palos Verdes Hills south of Los Angeles, California; the highest and oldest is about 1300 feet (396 m) above sea level.

Landforms of Marine Deposition

Sediment moved by longshore and beach drifting is laid down in places of low energy to form characteristic landforms. This sediment derives mostly from streams entering the sea, but it can derive from erosion of coasts as well. Occasionally, coarser sediment may be washed in shoreward by storm waves. Perhaps the most obvious and well-known landform of deposition is the **beach**—a shore of wave-washed sediment, extending from the low water line upward to the first land vegetation or a sea cliff. (Other definitions may be less encompassing.) Beaches include the upper *backshore*—upward from the high water line—and the intertidal, lower *foreshore,* which may contain ridgelike *bars* in its lower part. The word *beach* normally conjures up an image of a shore of sand—ideally white, dazzling sand as on west Florida beaches—but some beaches are of pebbles, cobbles, and even boulders. It depends on the available sediment and the wave energy. The beach and associated features change constantly. Beaches may move slowly landward as coasts undergo heavy wave attack, and migrate seaward as abundant sediment accumulates. And they characteristically lose sand during storms and regain it during more halcyon times. Since storms prevail during winter and spring and are less frequent during summer and fall, certain changes in a beach are seasonal. At La Jolla, California, for example, a beach is characteristically of sand in summer and fall, and of gravel—sand is removed by storms—in the winter and spring.

Man tries to stabilize beaches from longshore movement by constructing low walls (groins) at right angles to shore. But these don't work all that well. Some sand is trapped on the updrift side of the walls, but only temporarily. And shore erosion is the usual consequence on the downdrift side.

Along a fairly straight shoreline interrupted by bays or estuaries, longshore and beach drifting extend the beach and deposit sediment into the deeper water of these embayments to form a **spit** (Figure 4–4)—a ridge of sand or gravel projecting from land into open water. Where longshore movement—which is always toward the free end of a spit—is deflected landward, a curved spit or *hook* results. Sandy Hook, New Jersey and Cape Cod, Massachusetts are notable examples. Further extension of a spit blocks the mouth of a bay to form a **baymouth bar** or bay barrier. Active tidal currents may breach the bar in places or not allow it to form at all. If a spit extends outward and connects an island to the mainland or to another island, a **tombolo** is formed.

Elongate ridges of sand paralleling typically low-lying coasts—but separated from them by lagoons—are **barrier islands.** They may form by (1) breaching of spits by tidal inlets or

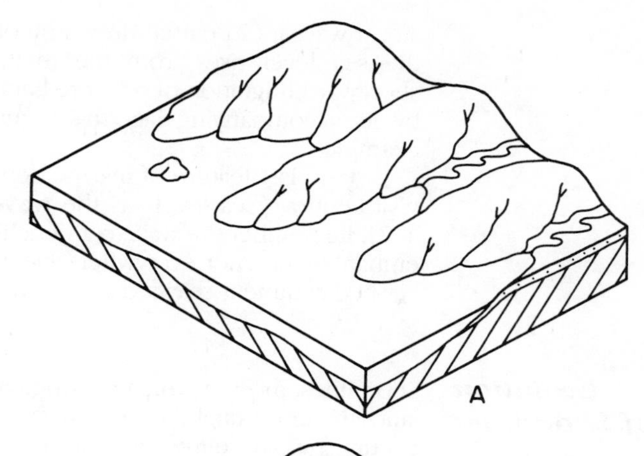

FIGURE 4–4. Evolution of shorelines.

A. A drowned shoreline is initially sculptured by stream erosion.

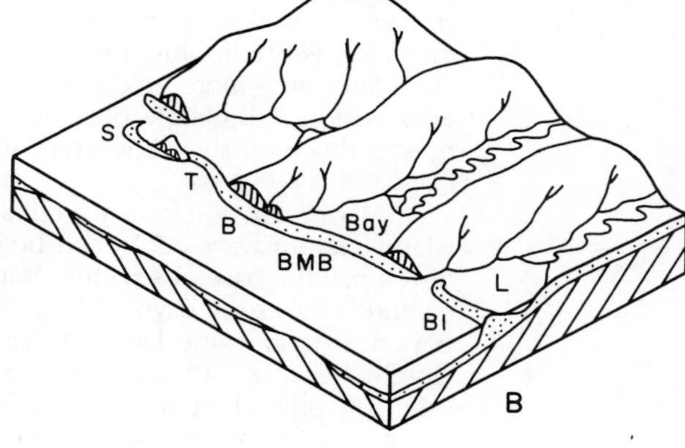

B. Marine erosion cuts sea cliffs on projecting headlands and islands, and beaches (*B*), spits (*S*), tombolos (*T*), baymouth bars (*BMB*), and barrier islands (*BI*) form. Bays and lagoons (*L*) fill with sediment.

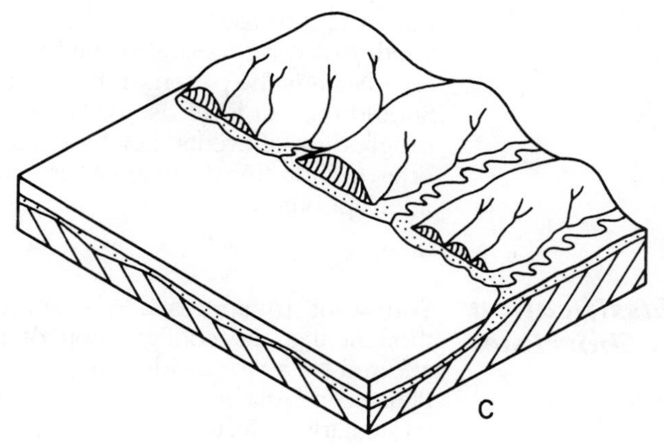

C. Planing of headlands and sealing off of bays and lagoons create a relatively straight shoreline. Retreat of the shoreline landward continues until the wave-cut bench is so wide that wave erosion essentially creases.

storm waves; (2) partial drowning of beach ridges-with-dunes as the sea level rises from the melting of glacial ice; and (3) shoreward migration of offshore bars. Padre Island, Texas, backed by its accompanying lagoon—Laguna Madre—is a well-known example.

One last feature of marine deposition—but one not readily visible above sea level—is the **wave-built terrace** (see Figure 4–2). Remember the wave-cut bench? The wave-built terrace is an embankment—not everywhere clearly distinguishable—of wave-washed sediment seaward of the wave-cut bench and surf zone.

Evolution of Shorelines

Shorelines, most of which are originally irregular, tend to become smooth and straight or gently curved with time. This change occurs by shoreline erosion and deposition performing in concert. Assuming no interruption by outside forces, shoreline evolution is predictable. But the *rate* at which it happens is not. Shorelines with more erodible rocks and those more exposed to wave action will clearly become straighter faster. Let's imagine, now, a drowned shoreline originally sculpted by stream erosion (Figure 4–4).

Marine erosion cuts sea cliffs on projecting headlands and islands, and a wave-cut bench begins to form in time. Beaches develop at the base of sea cliffs. With more sediment, beaches are extended into spits, baymouth bars, barrier islands, and tombolos. Bays seal to become lagoons that gradually fill with sediment, largely because of delta-building at their heads. Any islands originally present are wiped out. Planing of headlands and sealing off of bays, then, delineate a relatively straight shoreline. It will retreat farther landward until the wave-cut bench reaches such a width that wave erosion is inhibited or essentially eliminated.

Such is the pattern if the shoreline remains relatively stable. Should the sea level rise or fall, or the coast rise or subside, the progression of events sketched here would be interrupted. Rarely does a shoreline change progressively and completely without interruption.

Classification of Shorelines

You want to approach any shoreline classification with the thought that the configuration of many shorelines is markedly affected by a worldwide sea level rise caused by the melting of Pleistocene glaciers. Shoreline types 1, 3, and 4 (Table 4–1) particularly reflect this. A second point for your consideration is that more types of shorelines result from processes on land than from those occurring in the sea. In spite of the many shoreline types—not all of which are given in Table 4–1—this chapter dwells on only two, types 7 and 8, because landforms giving rise to nearly all of the other shorelines are covered in other chapters. Shorelines of predominantly marine erosion (type 7) (see Figures 4–2, 4–3) are common in Oregon and California, and those of

Table 4-1. CLASSIFICATION OF SHORELINES[1]

CHARACTERISTIC FEATURES	*LAND OR SEA PROCESSES PREDOMINATE*[2]	*EROSION OR DEPOSITION PREDOMINATES*[2]	*OTHER*	*NAME*
Drowned stream system	Land	Erosion	Valleys become bays and estuaries, divides become peninsulas and islands	1. Shoreline of stream erosion
Deltas, drowned alluvial fans	Land	Deposition		2. Shoreline of stream deposition
Drowned valley glacier system	Land	Erosion	Valleys become fiords, divides become peninsulas and islands	3. Shoreline of glacial erosion
No or partial stream system; low-lying highly irregular shoreline	Land	Deposition	Partly submerged moraines, drumlins, eskers, kames	4. Shoreline of glacial deposition
Cut lava flows, breached volcanic craters	Land	Erosion		5. Shoreline of volcan-ism[3]
Faults, folds	Land	Erosion	Fault-controlled shorelines straight; fold-controlled shorelines straight or irregular, with grain to eroded, folded beds	6. Shoreline of faulting or folding[3]
Sea cliffs, arches, caves and stacks; wave-cut benches (low tide)	Sea	Erosion	Uplift of shoreline implied in places	7. Shoreline of marine erosion
Beaches, spits, bars, barrier islands	Sea	Deposition	Much sediment available	8. Shoreline of marine deposition
Coral reefs, mangrove swamps, or marsh grass	Sea	Deposition		9. Shoreline of organisms

[1]Marine shorelines are implied in this table, but all but Shoreline 9 may occur in the larger lakes as well.

[2]Either predominates now or did so in the past.

[3]May be considered part of Shoreline 7.

mainly marine deposition (type 8) are frequent along the United States south Atlantic and Gulf coasts. For "Shorelines of Organisms" (type 9), keep in mind that coral reefs—as well as those of lime-secreting algae and other organisms—are of three main types: (1) *fringing* (attached to a landmass); (2) *barrier* (separated from a landmass by a lagoon, like a barrier island); and (3) *atoll* (generally ringlike with an enclosed, central lagoon). Reefs are usually submerged at high tide.

This classification, as with most others—be they of minerals or rocks and fossils—is not foolproof. On many shorelines shaped by marine, nonorganic processes, for example, *both* erosion and deposition occur. You must determine which of the two predominates to apply the classification. In other cases, you might require further information. In coastal Maine, for example, you might call a highly irregular shoreline—lacking a stream system—one of glacial deposition. In fact, though, knowing that scoured rock basins and finger lakes are present would reveal a shoreline of (continental) glacial erosion. In spite of its shortcomings, this classification should do you right in most cases, even only with the aid of maps and aerial photographs.

Selected Reading

SNEAD, R. E. *Coastal Landforms and Surface Features, A Photographic Atlas and Glossary.* Stroudsburg, Pennsylvania: Hutchinson Ross Publishing Company, 1982.

5 *WIND-RELATED LANDFORMS*

Unlike water and ice, wind lacks significant erosive power, and landforms of wind erosion are few. Most result from sand moved in arid or semiarid regions by wind and let down when the speed of the wind is checked. If wind-related landforms don't come to your mind readily, imagine yourself a central character in a desert movie, rocking along in a camel caravan that inches through a sand sea of undulating dunes under a blazing sun. The wind comes up and a sandstorm engulfs you. Much to your surprise, there on your camel you don't feel the stinging sand projectiles in your face. The reason will be touched upon later.

So the main wind-related landforms—dunes—occur in deserts. But other places—as sources of sand—produce dunes as well. Many shorelines, you might remember from Chapter 4, are reservoirs of sand derived mainly from streams entering the sea. Onshore winds shape this sand into dunes. Stream courses, with their channels and floodplains, provide sand for reworking by wind, as do areas of glacial outwash fashioned by melt water streams. And, whether in true deserts or not, areas of loosely cemented sandstone provide another sand source.

Sediment finer than sand—dust—is, of course, carried by wind in enormous amounts. Consider, for example, the great dust storms of the Great Plains during the dirty 1930s that caused bothersome fine sediment to infiltrate dwellings and organisms alike. But dust does not accumulate in conspicuous, easily recognizable landforms, and need not be dwelt upon in this chapter. The consideration of dust, though, leads to a worthwhile point. Although wind-related landforms prevail under an arid or semiarid climate, a subhumid or humid climate may accomodate them as well. It depends on a local or anomalous climatic situation. Consider the coastal dunes in relatively wet Oregon or Michigan, or the normally relatively humid climatic region of the easternmost Great Plains, succumbing occasionally to dust bowl conditions.

Work of the Wind

You can think of wind as a thin, easily moveable fluid; water is thicker, and, in Chapter 3, we considered glacial ice a thick, heavy fluid. Consequently, wind can carry only relatively small particles—mostly of sand size and smaller—and is more selective, therefore, of what it carries. Much of what geologists know about the wind's ability to pick up and move sand grains and the mechanics of this process was worked out by R. A. Bagnold and published in his classic book, *The Physics of Blown Sand and Desert Dunes* (1941).

Sand begins to move at a wind speed of about 11 miles per hour (18 km per hour) as the grains roll and slide. As the speed picks up, grains begin to jump into the air. Those landing on a hard rock surface or gravel particles continue to bounce along. Those striking other sand grains cause a splashing effect, each ejected grain striking one or more others, and the grain jumping spreads. Before long, a surface of sand is set in motion. Let's magnify the sand grains to the size of ping pong balls. Drop a single ball on the game table, and it bounces readily until its energy is spent. Now place a *layer* of balls on a part of the table. Throw a ball forcefully onto this layer, and watch the others jump and scatter. Most sand grains move by jumping, many more so than by rolling and sliding.

But the sand grains don't jump all that high. Most skip within a few inches of the ground and rarely above 6 feet (2 m). Sandblasting effects on poles, rocks, and other objects are seldom significant at a height of 1.5 feet (0.5 m). So people's heads and shoulders frequently project above sandstorms. (Their legs may be sandblasted, but their upper body parts are O.K.!) Now you see why I didn't share your surprise over the lack of stinging sand grains you felt while perched on your imaginary camel.

The wind's selective carrying ability causes the grains of most dune sand to fall in the range of 0.006 to 0.012 inch (0.15 to 0.30 mm), what geologists call *fine sand.* Other grades of sand are very fine, medium, coarse, and very coarse. So, you see, dune sand is near the fine end of the sand scale.

Dust particles move neither by rolling, sliding, nor by jumping, but are held *suspended* in turbulent, eddying air. Less wind speed is needed to move dust than sand, but dust particles may be difficult to lift initially. Why? Because the fine dust particles produce a smooth surface to the wind and are naturally cohesive. Usually some disturbance—such as off-the-road vehicles, animals, or jumping sand grains—is needed to allow wind to move dust readily.

Landforms of Erosion

Wind erodes by simply lifting and removing sediment as well as by abrading rock. The lifting and removal of sediment may scoop out a **blowout**—a hollow or depression—many feet to several miles in diameter. Geologists consider many depressions in the

United States Great Plains to be blowouts. How do you tell? Well, you might check for any alignment or elongation with prevailing wind direction, or look for flat-topped pedestals of sediment within the blowout that remain behind. Too, a residue lag gravel or **desert pavement** may line the floor of the basin, left behind as the wind removed the finer sand and dust. The resistant desert pavement and the water table (see Chapter 6) limit the depth of a blowout.

Abrasion is the sandblasting effect causing grooving, pitting, polishing, and faceting on large rock surfaces or gravel particles. Certain wind-sculpted rocks or **ventifacts**—not to be confused with human-fashioned rocks or artifacts—may have two or more flat surfaces or facets, formed as prevailing wind changes direction or a pebble, cobble, or boulder is turned over to expose a new surface to abrasive attack.

Landforms of Deposition

Dunes, mounds or ridges of sand, are the most prevalent of wind-deposited landforms. Many form about an obstacle—a rock or bush will do—that creates a *wind shadow* or pocket of quieter air in which eddies occur. The wind shadow, mostly downwind from the obstacle but to a lesser degree upwind, accumulates sand until a dune forms and creates its own wind shadow. A dune migrates as sand jumps, rolls, and slides up the gentler upwind slope, builds up at the crest, and slips or slides down the steeper downwind slope or *slip face* in thin, tonguelike masses. Slumping occurs when the downwind slope reaches about 34°, the angle of rest or maximum slope at which dry sand is stable. Slicing through a dune with a huge, imaginary knife would reveal most of the internal layering parallel to the slip face. Dunes resemble the sand ripples on stream beds, lake and sea bottoms—and on the backs of dunes—and other wave forms and bars. Quartz, a hard, durable mineral (see Chapter 14), makes up most dunes. But even a soft mineral like gypsum constitutes dunes, as at White Sands National Monument, New Mexico. Other materials include volcanic ash and even such heavy minerals as magnetite.

Common types of dunes are transverse, barchan, longitudinal, U-shaped, and star (Figure 5–1). They usually occur in groups and may merge and intergrade into complexes. Wind speed and constancy of direction, sand supply, and amount of vegetation primarily affect their formation (Figure 5–2). **Transverse dunes** (Figure 5–3)—at right angles or transverse to prevailing winds—form under low to moderate, constant winds moving an abundant supply of sand. Little or no vegetation is associated with these dunes, which prevail in deserts and along shorelines. Transverse dunes may be up to several thousand feet long and commonly up to about 30 feet (10 m) high. An area of large transverse dunes is the Namib Desert in southwestern Africa.

Barchan dunes (Figure 5–4), or simply barchans, are perhaps the most interesting. They are crescent-shaped dunes with

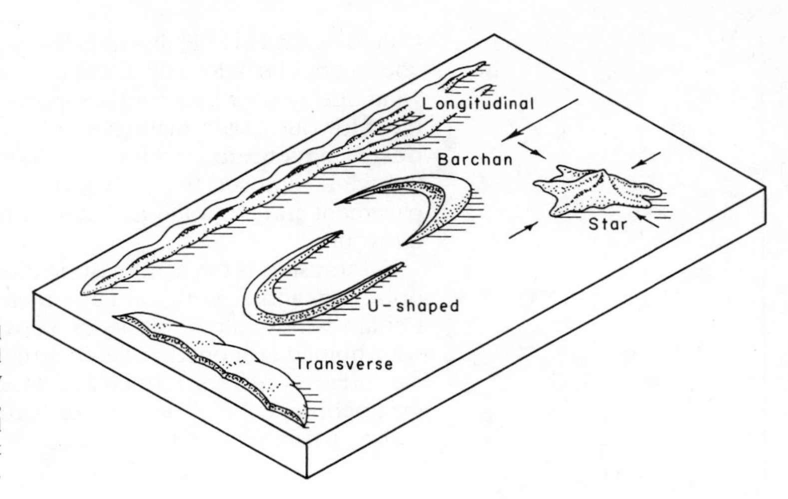

FIGURE 5–1. Main types of sand dunes. Arrows indicate wind direction; the single, long arrow relates to the longitudinal, barchan, U-shaped, and transverse dunes. Dunes are not drawn to relative scale.

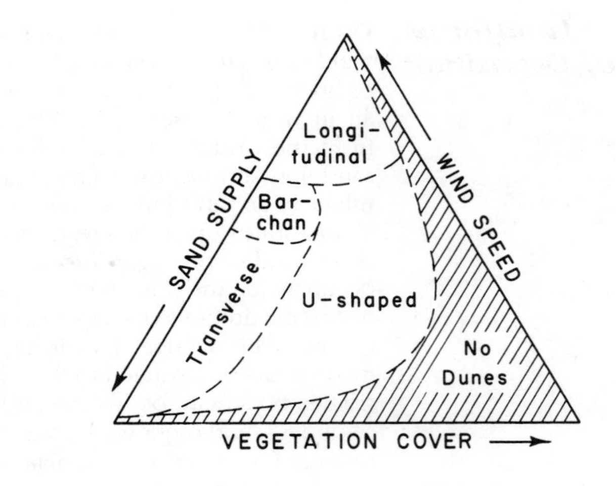

FIGURE 5–2. Relationship of four types of dunes to three factors affecting their formation. Arrows point in the direction of increase of each factor. A barchan, for example, forms under scanty sand supply and vegetation cover and moderate wind speed. (Modified from J. T. Hack, "Dunes of the Western Navajo Country," *Geographical Review,* 31 (1941), p. 260.)

FIGURE 5–3. Transverse dunes, Saudi Arabia. Slip faces toward the upper right indicate the wind blew from the lower left. (Courtesy of Aramco.)

FIGURE 5–4. Barchan dunes, with ripples on their surfaces. Shadowed slip faces on the right indicate the wind blew from the left. Along the Columbia River near Biggs, northern Oregon.
(Photograph 504 by G. K. Gilbert, U. S. Geological Survey.)

horns or tips that point downwind and curve around the slip face. They are often well developed on desert floors where moderate, constant winds move and shape a scanty supply of sand. Vegetation is scarce. Barchans are relatively small, up to about 1300 feet (400 m) wide; their height is generally about one-tenth of the width. They commonly migrate about 15 to 100 feet (5 to 30 m) per year, the smaller ones somewhat faster. Barchans may form at the margins of a transverse dune field where sand is limited. Vast areas of complex barchans occur in the Libyan Desert. Smaller areas of well-developed barchans include White Sands National Monument, New Mexico, and southeastern Washington.

Longitudinal dunes (Figure 5–5)—also known by the Arabic **seif,** meaning sword—are long, narrow, and parallel to prevailing winds. Actually, winds blow parallel at times, but at an angle at others as the dunes widen. Wind speed is high, but somewhat variable in direction, the sand supply is scanty, and the vegetation meager or lacking. Longitudinal dunes may reach several hundred feet in height—as in southern Saudi Arabia and eastern Iran—and several hundred miles in length. Extensive areas noted for longitudinal dunes are the Sahara and the large deserts of western Australia.

Now, let's pause for a moment and attempt to visualize a

FIGURE 5–5. Longitudinal dunes, Saudi Arabia. (Courtesy of Aramco.)

possible relationship between the three dune types just described. With your back to the wind, imagine your arms, extending straight out from your body, forming a transverse dune under a low to moderate wind and a good supply of sand. As the wind speeds up and sand diminishes, your arms swing forward forming a barchan. The wind blows harder still, and the barchan "smears out" into a longitudinal dune as your arms touch, fully extended forward (see Figure 5–2).

U-shaped, or **parabolic, dunes** somewhat resemble barchans, but with a slip face on the outside of the U, which is open upwind. Low to moderate, constant winds scoop blowouts in areas of considerable sand, particularly along shorelines, and U-shaped ridges—surrounding the blowout—become snagged by vegetation. Elongate U-shaped dunes assume a hairpinlike shape.

FIGURE 5-6. Star dune, Rub' al Khali region, southeastern Saudi Arabia. (Courtesy of Aramco.)

FIGURE 5–7. Sand "mountains," Rub' al Khali region, southeastern Saudi Arabia. (Courtesy of Aramco.)

Star dunes (Figure 5–6) are isolated, pyramidal sand peaks resembling a star in plan, with several sharp-crested ridges radiating from a central point. The wind blows from several directions and piles up sand against a central mass. Up to a few hundred feet high, they may occur at junctions of intersecting linear dunes. Star dunes are well displayed in southern Saudi Arabia and western Egypt.

Keep in mind that all dunes do not fit into one of the four categories we've just examined. Some are simply irregular hills or "mountains" (Figure 5–7) of sand. Too, when old dunes are completely covered by vegetation, they may be difficult to recognize (Figure 5–8).

Before we leave sand dunes, let me ask you this: Can dunes form in anything besides sand? Answer: yes—in *snow.* Most dunes we've examined can be generally recognized in snow. If you live or travel in snowy regions that are also blessed with the necessary ingredient, wind, look for them. Keep in mind the conditions necessary for the formation of each. So, if you really wish to see a *snow barchan,* look for it where a moderate wind is blowing scanty snow around. And watch for all sizes. Some snow barchans may be only a few to several feet wide.

FIGURE 5–8. Dunes covered by vegetation. Fence posts traversing dunes are visible in the upper left.

6 GROUND WATER–RELATED LANDFORMS

You are flying over central Florida. Your seatmate, next to a window, calls out, "Hey, look at all those lakes down there!" Leaning toward your companion, you glimpse the terrain below. Numerous, small lakes, all right—several of them even nearly circular. Your geological mind races for an explanation. "They're formed in jumbled sediment dumped by a glacier," you reply omnisciently, "some as insulated ice blocks melted out later and the little basins filled with water." Shortly after speaking, though, you sense you've committed a geological blunder. Glaciers in Florida? Come to think of it, too, there weren't any real hills between the lakes. This was flat country pitted with lakes. The circular ones resembled water-filled bomb craters. But no major battlefield of modern times here.

A day later you ask a Floridian geologist about the lakes, and you are told they are the work of ground water. Ground water? Having heard only of innocent ground water drawn from wells, you listen, transfixed, to the saga.

Ground Water

Ground water is the water in the spaces in sediment and rock beneath the earth's surface: in pores between rock and mineral grains, in fractures and cavities dissolved out of rock, and in gas-bubble cavities at the surfaces of lava flows (see Chapter 15). The greater the **porosity**—percentage of pore space—the more ground water a sediment or rock can hold. Most ground water originates as precipitation that infiltrates into porous soil, sediment, and rock.

Water seeping into the ground finds its way into one of two zones. In the upper, the **zone of aeration** (Figure 6–1), most pore space is filled with air, except, of course, directly after precipitation. You know, though, that soil and sediment and rock just beneath the surface tend to contain some moisture. Beneath is the **zone of saturation** where all pore space is saturated with water. At the boundary between the two, or at the top of the zone of saturation, is the **water table.** Generally, the water table reflects

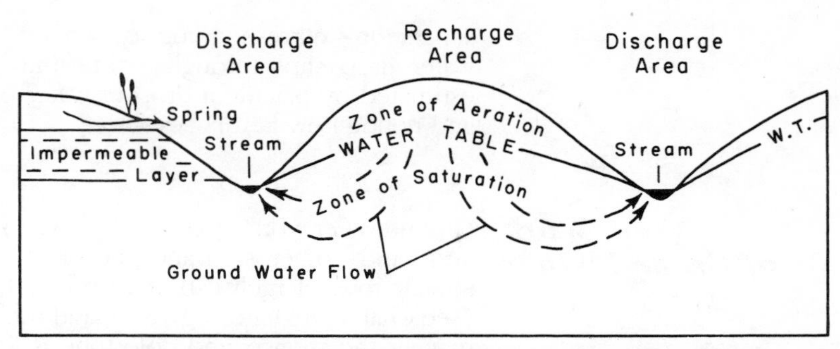

FIGURE 6–1. Ground water occurrence—mostly below the water table—and generalized flow in uniformly permeable rock material eroded by streams. Because of an impermeable rock layer—such as shale—a spring issues above the level of the water table. The water table may be flatter in limestone regions.

topography, higher under hills and lower under valleys or flat where the surface of the ground is flat. Springs are places where the water table intersects the surface, as are commonly the edges of perennial streams, lakes, and marshes. The water table rises and falls with wet and dry periods.

Ground water moves—once it reaches the zone of saturation—from areas of high water table to those of low, or from areas of high to low water pressure. Movement is along sweeping, curved paths—in the direction of the water table slope, but not directly down it—that trend downward as well as laterally. This is because a water table surface resembles a mobile, underground wave with water continually moving downward—by force of gravity—from the higher parts under greater pressure. The slow movement of ground water can be seen at sources of springs. Rates usually vary from about less than a foot per year to several hundred feet per day; only in caves and tunnels does ground water flow equal that in slow streams. On a large scale, we can think of a ground-water system, of higher *recharge* areas where water enters it and lower *discharge* areas where water leaves it after having percolated through.

Movement of ground water is largely dependent upon **permeability,** the ability of an earth material to *transmit* it. Rocks not only must have pore spaces, but they must be also connected. Sandstone and conglomerate commonly are highly permeable, providing pore spaces not filled with mineral matter. Seemingly "tight" rocks such as basalt and dense limestones (see Chapters 15 and 16) may be highly permeable if fractured. Shale, although highly porous, is usually poorly permeable because pores are too small to allow ready movement or are not connected. How the size of pore space affects permeability can be perceived by touching a thumb and forefinger together and placing a drop of water at their point of contact. Slowly separate the two. At first, the water drop is held suspended by the attraction of thumb and forefinger on it and by water particles on one another. Upon further separation—corresponding to increased size of pore space—the water drop slips through. Permeability—high or low—can be estimated by placing a drop of water on a broken rock surface and noting how fast it seeps in.

Work of Ground Water

Ground water works geologically—as do surface water, glacial ice, and wind—by erosion and deposition. It erodes mainly by dissolving rocks: largely carbonate rocks, rock gypsum, and rock salt (see Chapter 16). Rock gypsum and rock salt are relatively rare at or near the surface, and dolostone is less readily dissolved than limestone (or marble). It is limestone, then, that ground water primarily dissolves.

Rain water and ground water are slightly carbonated by carbon dioxide added from the air and soil. (Carbonated beverages get their "zip" and "fizz" from added carbon dioxide.) So part of the water is converted to carbonic acid:

Water + carbon dioxide = carbonic acid.

Nearly insoluble in water, limestone readily dissolves in carbonic acid:

Carbonic acid + limestone (calcite) = calcium bicarbonate (in solution) + water + carbon dioxide.

Deposition or precipitation of limy material also depends on carbon dioxide. As calcium bicarbonate-laden water drips or trickles in caves, carbon dioxide is released by water evaporating, or the gas is passed off as the water is agitated. Increased concentration of the limy matter finally causes it to precipitate from solution:

Calcium bicarbonate = carbonate dioxide (given off) + limestone (calcite; deposited) + water.

Landforms of Erosion

Most obvious erosional landforms are embodied within one word—**karst,** a solution-controlled landscape named after the limestone Karst district on the east side of the Adriatic Sea, primarily in western Yugoslavia. Other regions of well-developed karst include southern France, southeastern Asia, the Nullarbor area of southern Australia, northern Yucatan, Jamaica, Puerto Rico, central Florida, the Appalachian belt from Pennsylvania to Alabama, central Kentucky to southern Indiana, and the Ozark area of southern Missouri.

What is needed for karst to form? (1) Limestone or some other soluble rock. (2) Soluble rock with many fractures and closely spaced bedding surfaces along which acidic ground water concentrates its solution activity. Disseminated water moving

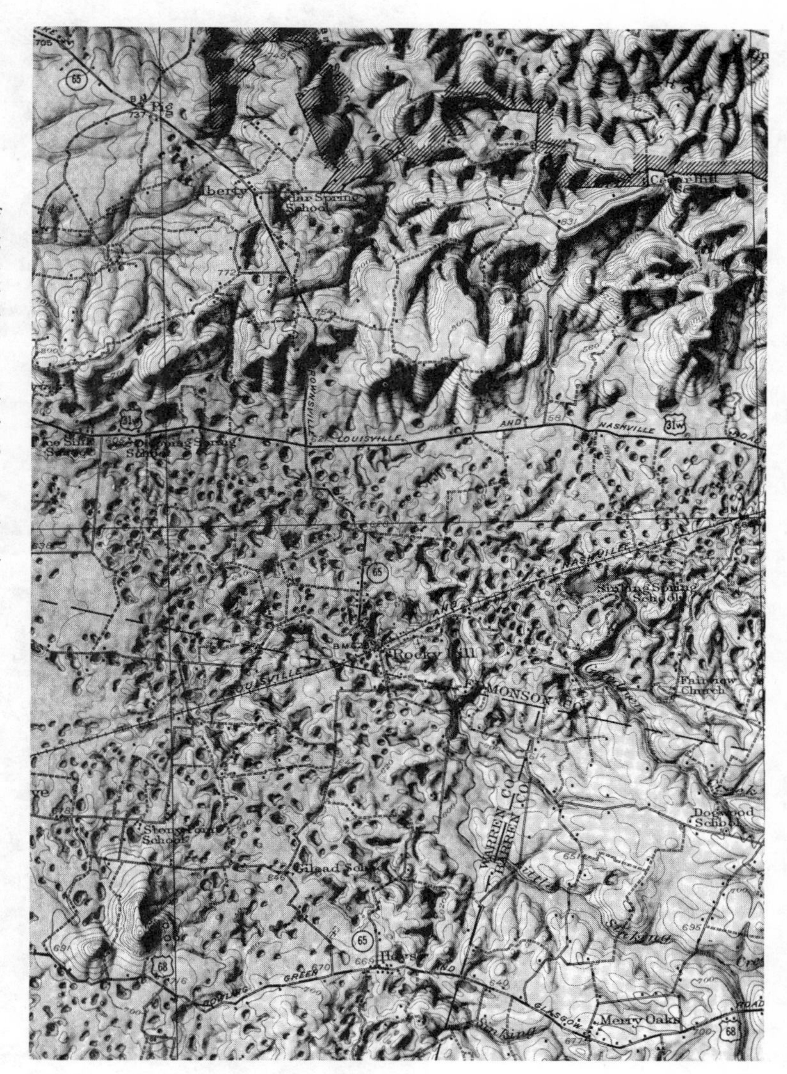

FIGURE 6–2. Shaded relief topographic map of karst showing abundant sinkholes and solution valleys—mostly in the lower part of the map. The southern boundary of Mammoth Cave National Park—diagonally ruled—is in the upper part of the map; southwestern Kentucky. The air distance between Rocky Hill (near center) and Liberty (upper left) is 3.8 miles (6.1 km). Rocky Hill is 8 miles (13 km) east-northeast of the 500-foot (152 m) sinkhole in Figure 1–3. (From U. S. Geological Survey Mammoth Cave Quadrangle, 1955.)

through homogeneously porous and permeable rock is apt to pass through without forming solution cavities. Although fractured, the rock requires sufficient strength so as not to collapse immediately after partly dissolving. (3) A humid to subhumid climate leading to at least moderate precipitation to provide a continuous replenishment of ground water. (4) Continually moving ground water that maintains solution. Stream valleys entrenched into soluble rocks allow ground water to percolate toward them in its journey from areas of a higher water table.

The most widespread of karst landforms are **sinkholes** (Figures 6–2, 6–3; see also Figure 1–3)—solution cavities open to the sky, formed by solution at the surface or by the collapse of

FIGURE 6–3. Sinkhole in Shelby County, central Alabama, the "December Giant:" 425 feet (130 m) long, 350 feet (107 m) wide, and 150 feet (46 m) deep. Slumping (see Chapter 7) is well developed. (Photograph 140 by U. S. Geological Survey; December 2, 1972.)

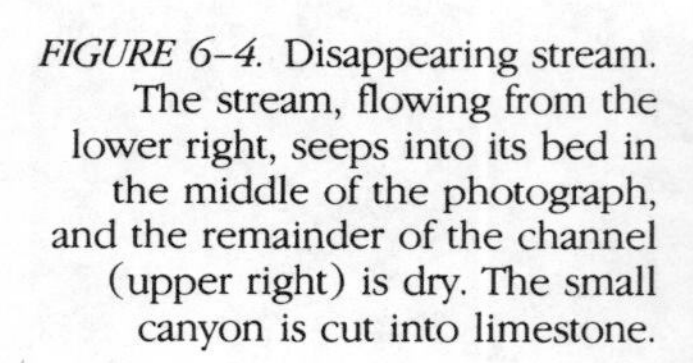

FIGURE 6–4. Disappearing stream. The stream, flowing from the lower right, seeps into its bed in the middle of the photograph, and the remainder of the channel (upper right) is dry. The small canyon is cut into limestone.

FIGURE 6–5. Karst in limestone dominated by conical hills in a tropical climate. Near Manati, Puerto Rico. (Photograph 371 by W. H. Monroe, U. S. Geological Survey.)

caves and **caverns** (roofed-over, underground solution cavities. Caverns are larger than caves). A few feet to several hundred feet in diameter, hundreds of sinkholes may occupy a square mile in some places. Throats of sinkholes may clog with mud, and sinkhole ponds and lakes result. Such water bodies may form, too, where the water table is high, as in central Florida. Sinkholes may merge and coalesce with others to form blind or closed **solution valleys.**

Surface water diverted into the subterranean solution system produces **sinking** or **disappearing streams** (Figure 6–4), which may follow tunnels for some distance and emerge to flow on the surface once again. Stream sediment in some caverns and natural tunnels attests to normal stream erosion playing a part, but most cavities in soluble rock seem to result from solution. The collapse of tunnels creates **natural bridges,** here-and-there remnants of those tunnels left standing. Such a bridge is 215–foot (66 m)-high Natural Bridge, near Lexington in west-central Virginia, crossed by U. S. Highway 11.

Let's briefly examine how a karst evolves in a temperate climate. Streams dissect a soluble rock terrain until a deep valley is cut. Ground water dissolves the rock along fractures and bedding surfaces as it moves toward the valley. Caves and caverns riddle the rock, and sinkholes, solution valleys, and sinking streams appear on the surface. In time, most of the original surface is destroyed, and a stream drainage system becomes obliterated. The soluble rock is finally dissolved away. Only then, when insoluble rock appears once again at the surface, does a stream drainage system reappear. In the tropics, solution occurs mostly at the surface, and is insignificant at depth. Consequently, conelike and towerlike hills (Figure 6–5) prevail over sinkholes by the time the original surface is destroyed.

FIGURE 6–6. Stalactites (hanging from the ceiling), stalagmites (projecting above the cavern floor), and columns (fused stalactites and stalagmites), Carlsbad Caverns National Park, southeastern New Mexico. (Photograph 150-211 by the National Park Service.)

FIGURE 6–7. Siliceous sinter cone of Castle Geyser, once a quiet hot spring. Yellowstone National Park, northwestern Wyoming.

FIGURE 6–8. Terraces of limy travertine (see Chapter 16), Mammoth Hot Springs, Yellowstone National Park, northwestern Wyoming.

Landforms of Deposition

Let's assume that limestone caves and caverns are excavated by acidic water below the water table. If the water table is lowered, as by uplift or ground water–draining valleys cut more deeply, air-filled caves and caverns become sites of limestone deposition. Dripping water lays down **dripstone** as water evaporates and releases carbon dioxide: iciclelike **stalactites** hanging from the ceiling (*c* in the word can remind you of ceiling) and inverted iciclelike **stalagmites** (*g* in the word for ground) projecting above the ground or floor. Stalactites and stalagmites fuse to form **columns** (Figure 6–6). Flowing or trickling water lays down **flowstone** in the form of rock tapestries, ribbons, and falls. Both dripstone and flowstone are types of travertine (see Chapter 16). Among the famous caves and caverns with abundant and varied limestone deposits are Mammoth Cave, Kentucky and Carlsbad Caverns, New Mexico.

In areas of present or relatively recent igneous activity—such as Yellowstone National Park, Iceland, and New Zealand—ground water issues to the surface as **hot springs** and **geysers,** which are hot springs that erupt periodically. At the sites of these thermal features, ground water-deposited rock material builds mounds, cones (Figure 6–7), and terraces upon evaporation, cooling, loss of gas, drop in pressure, or precipitation by algae. Where the source rock is rich in silica, as is the felsite rhyolite, a deposit of porous **siliceous sinter** (see Chapter 16) results. Hot ground water passing through limestone lays down porous travertine, as in the terraces at Mammoth Hot Springs (Figure 6–8) in Yellowstone National Park.

Selected Reading

SWEETING, M. M. *Karst Landforms.* New York: Columbia University Press, 1973.

7 LANDSLIDE-RELATED LANDFORMS

It's a beautiful, still, mid-August night. The moon is full. But you can't sleep, although you're snug in your bedroll near the bottom of an impressive canyon. Your sister sleeps soundly nearby, and your parents have doused the lights in their house trailer. You glance at your watch—about 11:35 P.M.—and concentrate on sleep. A few minutes later, vibrating ground beneath jars you fully awake. You sit up with a start. Trees around you are whipping back and forth. Is this real, or are you having a nightmare? Above the thrashing of the whipping trees you hear a thunderous roar—and both are occurring during a still, clear night.

As your alarmed parents dash from the trailer, a powerful blast of air slams into the campground. Your father, grasping a tree, is strung out flag-wise before he is forced to let go. Buffeted by the air blast, your body is pummeled by trees and rocks as you are hurled along. A sharp pain pierces your left leg. The clear night turns dark.

In the morning only you—bruised and with a broken leg—and your mother survive from your family. Your father and sister share the same fate as 24 other campers in the canyon.

Sounds unreal? Could it happen? Well, it did, on August 17, 1959. At 11:37 P.M., the Hebgen Lake earthquake, centered a few miles northwest of Yellowstone National Park, triggered the Madison landslide (Figure 7–1) 17 miles (27.4 km) to the west on the south wall of the canyon cut by the Madison River. Nearly 40 million cubic yards (31 million cubic meters) of rock debris slid down the south wall of the canyon, burying a mile (1.6 km) of river and highway up to a depth of 220 feet (67 m). This slide mass dammed the Madison River, creating a lake that, three weeks later, extended upstream 6 miles (9.6 km) and reached a depth of 190 feet (58 m). Slide debris traveled up to a mile (1.6 km) and forced its way to 430 feet (131 m) above the riverbed on the north canyon wall. One block that rode the slide mass down is house-sized, almost 30 feet (9 m) on a side (Figure 7–2). The air

FIGURE 7–1. Rock debris of the Madison slide that dammed the Madison River and formed Earthquake Lake—in the background. The slide scar is off the view to the right. Rock debris forced its way to 430 feet (131 m) above the riverbed on the opposite (north) canyon wall (left). Madison County, southwestern Montana. (Photograph 216 by J. R. Stacy, U. S. Geological Survey.)

FIGURE 7–2. Dolostone block moved to the opposite (north) canyon wall by the Madison slide. Height of the person is 5.5 feet (1.7 m).

blast mentioned in our story—and reported by several eyewitnesses—was most likely caused by rapidly expelled air initially trapped by the descending slide mass. Some believe such trapped air provides a cushion or lubricant for the slide mass to move farther than it could without it.

What caused the Madison slide, one of the three largest rapid landslides to occur in North America within historic time? (The other two were the Gros Ventre slide near Grand Teton National Park in 1925 and the Turtle Mountain slide near Frank, Alberta in 1903.) An earthquake set it off, to be sure, but suitable geologic conditions allowed the event to happen. On the south canyon wall, deeply weathered, highly sheared gneiss and schist (see

Chapter 17)—with stress-induced layering and shear zones inclined steeply toward the canyon—were literally waiting for dislodgment. Fractured dolostone was also inclined toward the canyon.

Now, let's step back and examine the conditions that enhance the downslope movement of rock and rock debris generally. Here, I'll use *debris* to mean any loose rock material blanketing solid rock, including soil and sediment. Gravity, of course, is the driving force. It becomes effective when the internal resistance of any earth material to move downslope is overcome, by say, over-steepening of a slope. Oversteepening may occur by stream downcutting, wave erosion along shorelines, or human excavation. Landslides along steep highway cuts are notorious. Sudden shocks, as from earthquakes—this was seen in the Madison slide—blasting, or sonic booms of aircraft may, too, overcome an earth material's resistance to movement. Water enhances movement by adding weight and, especially when added to clayey materials, creates a lubricated surface upon which overlying materials may move. Water may partially saturate earth materials so they may flow as well as slide or fall. After periods of heavy rains or snow melt, then, steep slopes are particularly prone to landsliding. Lack or scarcity of vegetation encourages downslope movement. And fractures and layering inclined downslope—as in the Madison slide—do so as well.

Kinds of Landslides and Related Landforms

Downslope movement of rock and rock debris can be considered fast or slow: fast if you can readily see it, slow if you cannot. Geologists don't always agree as to what types of downslope movement should be considered landslides. Some say landslides are only the downward sliding or falling of relatively dry rock or debris; others would include flows, wet or dry. Since flowage may accompany sliding, for example, I am including all types of fast downslope movement within the category of landslides (Figure 7-3).

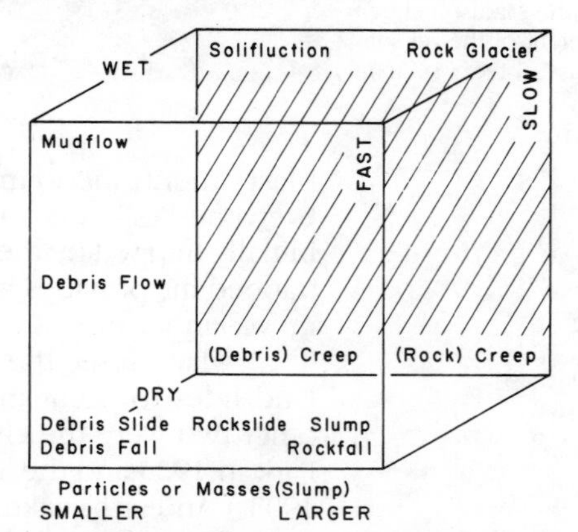

FIGURE 7-3. Kinds of landslides (*front panel*) and related features. (Modified from a diagram by J. R. Reid.)

Rockfalls are landslides wherein rock falls freely from cliffs or steep slopes. Rolling, bounding, and ricocheting may also be involved. **Debris falls** are the debris counterparts of rockfalls. Rockfalls and debris falls form accumulations of rock and debris at the base of slopes—**talus** (Figure 7–4)—that tend to resemble half-cones.

Rockslides, the most catastrophic of landslides, are rapid slides of rock along surfaces of weakness, such as layering or fractures. The Madison, Gros Ventre (Figure 7–5), and Turtle Mountain (Figure 7–6) landslides are all rockslides. In the Gros Ventre slide, the Gros Ventre River had oversteepened the lower slopes. Heavy rains and melting snow saturated sandstone and underlying shale, both inclined towards the river valley. The heavy, water-saturated sandstone simply slid down on the greasy shale. **Debris slides,** debris counterparts of rockslides, may roll or slide. A snow avalanche is similar to a debris slide. Both rockslides and debris slides produce bare slide scars and uneven slide masses. In slide masses

FIGURE 7–4. Talus. Note the dead tree in the foreground brought down with the rock debris.

FIGURE 7–5. Gros Ventre slide, near Grand Teton National Park, northwestern Wyoming. The relatively bare slide scar is evident, below which is uneven, vegetated rock debris. This slide dammed the Gros Ventre River and formed a lake similar to that of the Madison slide (Figure 7–1).

FIGURE 7–6. Turtle Mountain slide, Frank, southern Alberta. This slide moved along steeply inclined fractures in limestone that overlies weak shale, siltstone, and coal; it killed 66 people in the town of Frank. Alberta Highway 3 and the Canadian Pacific Railway cross the slide mass. (Aerial photograph T31L-214 © 1944 Her Majesty the Queen in Right of Canada, reproduced with permission from the National Air Photo Library, Department of Energy, Mines and Resources, Canada.)

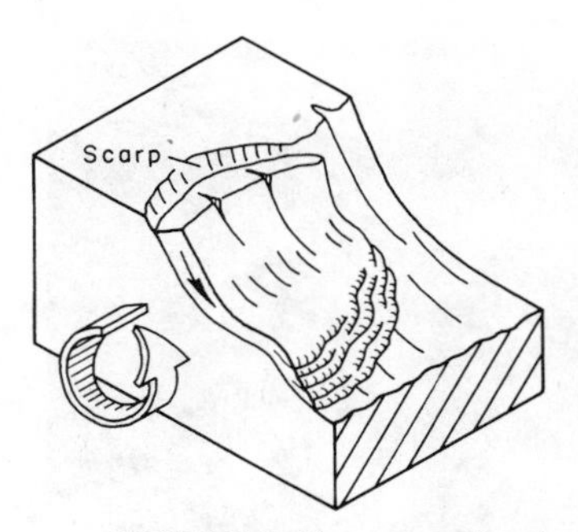

FIGURE 7–7. Slump associated with a debris flow. Note the scarp and the backward rotation of the slump mass.

of debris slides, particularly, hills or hummocks and ridges are separated by intervening depressions, resembling a topography similar to that of certain glacial moraines (see Chapter 3). **Slumps** (Figures 7–7, 7–8; see also Figure 6–3) differ from rockslides and debris slides in that blocks or masses of rock or debris move as discrete units, usually with backward rotation on curved surfaces down and away from scarps or cliffs from which they descend. Ponds or lakes may develop in the elongate depressions back of and between rotated blocks. Below the scarps rather chaotic topography may occur, as in rockslides and debris slides. Here, nearly parallel arclike ridges may alternate with similarly shaped depressions (Figure 7–9), especially where water-saturated materials may flow in the lower parts of slumps. Heavy rain may initiate slumping or contribute toward further movement once it takes place.

Debris flows and mudflows form where rock debris becomes water-soaked. **Debris flows** may have the consistency of fresh cement or a thick fluid; many occur in the lower parts of slumps. **Mudflows** are more fluid than debris flows, contain a high percentage of mud, and follow stream valleys or channels, spilling out into lobes or fans upon reaching low-lying terrain. They prevail in dry regions with scarce vegetation, generated by heavy, though infrequent, rains. Their high density allows them to carry buildings as well as boulders weighing many tons. Mudflows also move volcanic debris along the flanks of volcanoes. Flows of both types leave scars bare of soil and vegetation and lobelike ridges or corrugations at their fronts.

As a parting thought for this section on rapid downslope movements, be aware that falls, slides, and flows may all occur underwater, even though the resultant landforms are not generally visible.

FIGURE 7–8. Slump, showing bare scarp (cliff at upper part of photograph) and vegetated slump mass, slightly rotated backward.

FIGURE 7–9. Landslide topography—mostly from slumping—of near-parallel arclike ridges and intervening depressions, caused by stream (Peace River) downcutting and slope failure. Near the town of Peace River, west-central Alberta, Canada. (Aerial photograph A21819-40 © 1970 Her Majesty the Queen in Right of Canada, reproduced with permission from the National Air Photo Library, Department of Energy, Mines and Resources, Canada.)

Slow downslope movements are fewer than rapid ones. **Creep** is the extremely slow downslope movement—a fraction of an inch to perhaps a few inches per year—of rock, soil, and sediment. It prevails on smooth vegetated slopes in temperate and tropical climates. In exposures, tilted layers are seen curved downslope within the weathered zone (see Figure 17–2) and tree roots trail behind their trunks. On the surface, lower parts of tree trunks are bent downslope, and posts, poles, and gravestones are tilted; roads may be somewhat slung downslope. In areas of periodic freezing, frost heaving contributes significantly to creep. With freezing, particles are forced upward at right angles to the slope; upon thawing, such particles are let down vertically. Repeated freezing and thawing, then, passes particles downslope in small increments, slowly but constantly. Organisms also contribute to creep: Burrowing animals pile dug-out material on the downslope side of their burrows; large grazers following paths push debris downslope with their hooves; and plants wedge loose material downslope by their roots. Creep may produce lobelike bulges on hillsides.

To some a kind of creep and to others flowage, is **solifluction** (from the Latin *solum,* "soil," and *fluctuare,* "to flow"), a slow downslope movement of water-saturated debris. Most pronounced in polar or high altitude areas of permanently frozen ground, movement occurs during periods of thaw. During warm periods, melt water saturates thawed debris to a shallow depth that flows over a permanently frozen substrate, even on gentle slopes. Solifluction forms lobes or terracelike landforms on slopes.

Rock glaciers are tongues or streams of angular rock fragments at the base of cliffs or in valleys of cold, mountainous regions. Flowage is indicated by steep, lobed fronts and parallel, arclike ridges on their surfaces. Movement seems to take place largely by the flow of ice in the spaces between rock fragments. Rock glaciers, active in such places as Alaska, closely resemble ice glaciers, and some may be highly debris-covered remnants of former ice glaciers.

Related to downslope, gravity-induced movement of rock and debris is **subsidence,** the directly downward rapid or slow movement or sinking of parts of the earth's crust. Most subsidence near the earth's surface is caused by the removal of underground fluid—especially oil and ground water—and the collapse of underground cavities. Overwithdrawal of oil and water from pore space in rock and sediment leaves these earth materials unsupported; they compact, subside, and produce depressions or basins at the surface. In the San Joaquin Valley of California considerable subsidence has resulted from overwithdrawal of ground water. Collapse over naturally dissolved limestone, rock salt, and rock gypsum forms sinkholes (see Figure 6–3). Similar features form where people pump water underground to mine salt or sulfur. Salt-laden water is pumped back to the surface, and unsupported cavity roofs subside. Subsidence—forming pits, commonly elongate and aligned—is also common in regions of underground mining, as in the coal-bearing Appalachian region of the eastern United States. And volcanic craters and the larger calderas (see Figure 8–2) may, at times, form by subsidence as molten rock is withdrawn.

Selected Readings

HADLEY, J. B. "Landslides and Related Phenomena Accompanying the Hebgen Lake Earthquake of August 17, 1959," *U. S. Geological Survey Professional Paper 435-K* (1964), 107–138.

SHARPE, C. F. *Landslides and Related Phenomena.* New York: Columbia University Press, 1938.

WITKIND, I. J. "Events on the Night of August 17, 1959—the Human Story," *U. S. Geological Survey Professional Paper 435-A* (1964), 1–4.

8 VOLCANIC LANDFORMS

Volcanoes and lava flows are familiar to most of us even though we may not have seen them directly. Who in North America, for example, has been unaware of Mt. St. Helens in southwestern Washington since its historic eruption of May 18, 1980? In this chapter we will concern ourselves mostly with recognizing various types of volcanoes and lava flow-related landforms from a distance. For information on the rocks of volcanic landforms and igneous features, you might wish to check Chapter 15.

Volcanoes

Volcanoes are hills or mountains built up of molten lava, fragmented rock, or both, spewed from vents or fractures. At their tops are circular or elliptical depressions—**craters,** if less than a mile (0.6 km) in diameter, or **calderas,** if larger. Both craters and calderas form by explosion, collapse, or both processes. More than 500 volcanoes are considered active today, and most occur in narrow belts, especially within the circum-Pacific belt, the so-called Ring of Fire. It extends from southern Chile northward sporadically along the western edges of the Americas; from the Aleutian Islands it continues to Japan, the Philippines, the East Indies, and New Zealand. Volcanoes are scattered, too, within the Pacific, as in the Hawaiian Islands. Another major belt trends east-west from the Mediterranean region through southern Asia and joins the East Indies. The mid-Atlantic presumably has numerous volcanoes, but they become visible only at such places as Iceland, the Azores, and Tristan da Cunha. Volcano belts correspond, too, to the major earthquake belts. Earthquakes, in fact, are frequently monitored to help predict volcanic eruptions. The belts of closely associated volcanoes and earthquakes are believed to delineate the boundaries of constantly moving, major earth plates (see Chapter 12).

Volcanoes present different profiles (Figure 8–1) because of the materials of which they are constructed. **Shield volcanoes** (Figure 8–2) are flat domes—slopes are generally less than 10°—

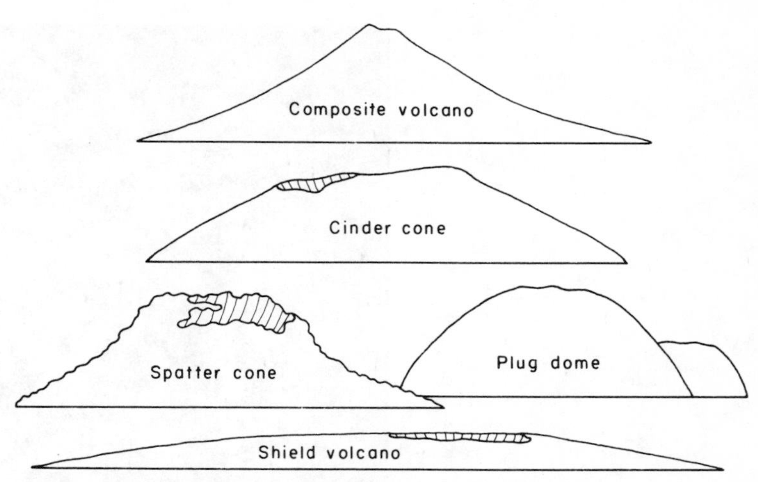

FIGURE 8–1. Profiles of major types of volcanoes. Profiles are not drawn to scale; shield and composite volcanoes are the largest, others are relatively smaller. Diagonal ruling signifies craters or a caldera (shield type).

FIGURE 8–2. Caldera of the shield volcano Kilauea, Hawaii, Hawaiian Islands. View is northward with Mauna Kea, another shield volcano, in the background; an extensive lava field separates the two volcanoes. Within the caldera, bordered by fault scarps, are the Halemaumau "Fire Pit" or crater near its southwest edge and lava flows on its floor, one of which has oozed out of the caldera (bottom of the photograph). A spatter cone, barely discernible, occurs within the Fire Pit. (U. S. Geological Survey photograph Hawaii 8 A.)

FIGURE 8–3. Cinder cone (top) and associated lava flow. SP Mountain, 26 miles (42 km) north of Flagstaff, north-central Arizona. (U. S. Geological Survey Photograph 1-35 GS-VVP.)

that resemble a warrior's shield in profile. They are composed mostly of lava of silica-poor basalt, hot—about 1600 to 2200°F (900 to 1200°C)—and thin or free-flowing, which results in gentle slopes. Although lava usually pours quietly from the crater and from fractures along a volcano's flanks, gushing lava fountains do occur. Hawaiian volcanoes, such as Mauna Loa, and several in Iceland are of the shield type. Mauna Loa is the world's largest known active volcano, many hundreds of square miles in area; it rises about 30,000 feet (9140 m) above the sea floor. This may seem huge, but the largest shield volcano on Mars—Olympus Mons—is twice as large and towers above surrounding plains. Calderas are common at the summits of shield volcanoes, formed largely by collapse as lava is drained temporarily from subterranean chambers. Gas-charged, basaltic, lava fountains may build up rough, steep-sided **spatter cones,** of spatter or blobs of lava. Generally small, they may reach a few thousand feet high.

Silica-rich lava similar in composition to rhyolite or andesite is cooler and thicker, and flows less readily. Some barely flows at

FIGURE 8–4. Dormant composite volcano with lava flows at the base, Mt. Shasta, northern California. (Photograph 3668 by J. S. Shelton.)

all, and heaps up into extremely steep-sided, bulbous **plug domes** over vents. Plug domes are similar in size to spatter cones. An area of plug domes is Mono Craters in eastern California.

The explosive accumulation of volcanic material—mostly volcanic ash, with coarser fragments—about a vent forms a **cinder cone** (Figure 8–3). Cone slopes are generally 30 to 35°, at the angle of rest of loose volcanic fragments. Cinder cones often group in clusters, and basalt flows may extend from their bases. A famous cinder cone is Parícutin, born in a western Mexican cornfield in 1943 and ceasing activity nine years later. At the close of eruptions it rose 1345 feet (410 m) high, reasonably large for a cinder cone.

Constructed of a combination of lava and fragmental material, **composite volcanoes** (Figure 8–4) characteristically display slopes of about 30° close to summits and 5° near their bases. Their materials, reflecting alternating periods of explosive and quiet eruptions, most often have the composition of andesite but range from rhyolite to basalt. Most of the world's larger volcanoes are of this type; examples include Mayon in the Philippines, Fujiyama in Japan, El Misti in Peru, Merapi in Java, and Mts. St. Helens and Shasta in the United States Cascades. Two-thousand-foot (600 m)-deep Crater Lake (Figure 8–5) in the Oregon Cascades occupies a caldera within a former composite volcano. Back in the Pleistocene, valley glaciers streamed down the slopes of this peak. Eruptions and collapse hollowed out the famous caldera. Later, a cinder cone—Wizard Island—with a lava flow issuing from its base appeared.

FIGURE 8–5. Caldera within a former composite volcano, Crater Lake, Crater Lake National Park, southwestern Oregon. Wizard Island, a cinder cone with a lava flow at its base, is in the foreground.

Lava Flows and Fields

Lava flows are outpourings or masses of molten lava or the solidified rock equivalent of that lava. Flows issue from cracks or vents. Basaltic lavas with little gas move relatively slowly and form flows many feet thick. The surfaces of such flows are of a rough, jumbled mass of angular blocks and cinders. The Hawaiian term for such flows is *aa* (AH-ah) (Figure 8–6). *Pahoehoe* (Pah-HOE-ay-hoe-ay) flows, of lava with considerable gas, are thinner, with wrinkled, ropy surfaces. As mentioned in the discussion of plug domes, rhyolitic or andesitic lavas are stiff and pasty, move more slowly, and so produce flows with steeper sides.

Flows have lobate margins and develop characteristic landforms on their surfaces. **Pressure ridges,** usually at right angles to the direction of flow movement, form by squeezing of the semirigid surface as the interior remains mobile. Cracks frequently adorn the crests of ridges, and lava may issue secondarily from them. **Squeeze-ups** are essentially moundlike equivalents of pressure ridges. Where basaltic lava is neither too stiff nor too fluid and flows with a moderate gradient, it may be confined in places to **lava channels**—open at the top—or roofed-over **lava tubes** or **tunnels;** both may extend for several miles. Drained, cooled tubes lure the inquisitive cave explorer. Some are well insulated and retain ice the whole year. The collapse of tubes creates elongate depressions resembling channels; short remnants produce natural arches or bridges.

Broad areas of lava flows, frequently with groups of cinder or spatter cones, are called **lava fields** (Figure 8–2) or **plains.** Such areas, displaying all of the flow-associated landforms mentioned here, are extensive within the Snake River region of southern Idaho.

FIGURE 8–6. Rough, jagged *aa* lava flow with a cinder cone in the distance. Craters of the Moon National Monument, south-central Idaho.

Lava fields are prevalent on the moon and Mars. The dark, lowland areas of the moon—*maria*—represent "seas" of basalt, extruded after the development of the light, heavily cratered highlands or *terrae.* Flow margins are not always obvious, presumably because of the extreme fluidity of the lava. Sinuous or winding rilles on the moon, and some on Mars, may be lava channels or collapsed lava tubes.

Evolution of Volcanic Terrains

After the extrusion of volcanic products, a terrain undergoes progressive change, providing renewed volcanism does not interrupt such change. Under light volcanism, lava flows follow stream valleys and dam streams. Diverted water runs along the sides of the more resistant lava flows and concentrates erosion there so as to create new valleys in those places. Eventually, the areas of lava flows become elevated, elongate, twisting tablelands—no longer resembling stream valleys—as the surrounding, more erodible rock wastes downward and away. These tablelands, in time, succumb and degenerate to separated mesas and buttes. Those volcanoes largely or exclusively of fragmented volcanic material erode rapidly to skeletal volcanic necks (see Figure 15–14) of more resistant rocks.

Under heavy volcanism, with emplacement of many lava flows, a landscape may be totally buried. Drainage, once again, is diverted, this time around the margins of lava fields. Some streams with steeper gradients, however, may cut through the lava pile. Ultimately, the more resistant, coalesced lava fields become an elevated lava plateau as surrounding softer rocks are worn down. One such plateau is the Columbia-Snake River Plateau (Figure 8–7), mostly in Oregon, Washington, and Idaho, where the basalt

FIGURE 8–7. Stacked basalt flows within the Columbia-Snake River Plateau, downstream from Palouse Falls, southeastern Washington. Right-angled bends in the stream are controlled by rock fractures. Wind-blown silt overlies the basalt in the right background. (Photograph 29 by F. O. Jones, U. S. Geological Survey.)

flow pile exceeds 10,000 feet (3000 m); emplacement began in the Miocene. Another lava plateau is the Deccan Plateau in India. Lava plateaus are cut further into mesas and buttes, which, too, eventually are destroyed.

In the sea, violent wave action consumes volcanic materials, the fragmented material first. Outright violent explosions, generated by entrapped steam, may blow away entire volcanoes catastrophically.

Selected Readings

BULLARD, F. M. *Volcanoes of the Earth.* Austin: University of Texas Press, 1976.

FOXWORTHY, B. L., and MARY HILL. "Volcanic Eruptions of 1980 at Mount St. Helens: The First 100 Days," *U. S. Geological Survey Professional Paper 1249* (1982), 1–125.

LAMBERT, M. B. *Volcanoes.* Seattle: University of Washington Press, 1980.

MACDONALD, G. A. *Volcanoes.* Englewood Cliffs, N. J.: Prentice-Hall, Inc., 1972.

SCHMIDT, R. G. *Atlas of volcanic phenomena.* Washington, D. C.: U. S. Geological Survey, 1974? (20 color charts)

9 ROCK DEFORMATION—RELATED LANDFORMS

The earth's crust is continually subjected to stresses; earth tremors—other than those associated with volcanic eruptions and similar localized events—reflect the crust's yielding to these stresses as rock masses jerkily grind along large-scale fractures. Old fractures signify similar crustal breakage in the past. But many rocks are also tilted and bent or folded into a variety of configurations. So rocks can yield to stress by bending and folding as well as by breaking. Rocks *bend?* In Chapter 3, we learned that ice, commonly thought of as brittle, can *flow* in glaciers. And glacial ice can be considered a rock of largely a single mineral. Yes, rocks bend when under high pressures and temperatures, both of which are found at considerable depths.

Landforms related to rock deformation of layered rocks are largely linear ridges and intervening valleys, molded by **differential erosion** or the faster erosion of some rocks than others. The differential rate varies with the degree of lithification, the minerals present, and the climate. Of the sedimentary rocks, sandstone, conglomerate, and limestone produce ridges, and shale forms valleys, if within an arid climate. In a humid climate, both limestone and shale characterize the valleys.

Dip and Strike

Imagine walking along the rocky, southern shoreline of Lake Superior in northern Michigan. You notice a sandstone bed tilted toward the lake, to the north-northwest. This tilt or inclination of the bed is its **dip,** estimated or measured (see Chapter 20) from a horizontal surface, in this case the water level. What is the greatest dip a tilted layer can have? Answer: 90° or a vertical right angle. The trend or **strike** of the bed is delineated precisely by the horizontal water line meeting it. Strike here is east-northeast, and is always at right angles to dip. Geologists can measure and orient any inclined bed or fracture in space by means of strike and dip. These values are necessary in projecting inclined beds and fractures beyond where they are visible.

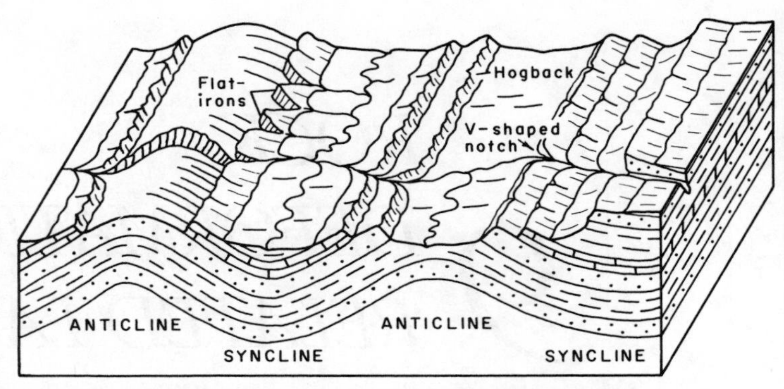

FIGURE 9–1. Eroded anticlines and synclines and resultant landforms. V-shaped notches—seen looking down on the eroded surface—point in the direction of dipping beds. Rock symbols in the front and right panels are: dots = sandstone, short lines = shale, and bricklike pattern = limestone.

Fold-controlled Landforms

Monoclines (Greek *monos,* "single," plus Latin *clinare,* "slope") are the simplest of folds, single steplike bends in flat-lying or gently dipping beds. **Anticlines** (Figure 9–1) are upfolds or uparched folds. Place several sheets of paper on a table or desk. Push opposite edges of the sheets toward each other with both hands. A buckling upward of the paper creates an anticline, each sheet corresponding to a folded rock layer. The sides or flanks of anticlines dip away from one another. **Domes** (Figure 9–2) are basically anticlines in which the beds dip outward more or less equally in all directions; many geologists, however, include rather elongate folds within domes. A well-known elongate dome is the Black Hills in western South Dakota and northeastern Wyoming. About 130 by 60 miles (209 by 96 km), it trends north-northwest; asymmetrical, strata dip more steeply on the east side. The higher exposed core of Precambrian intrusive igneous and metamorphic rocks is flanked by progressively younger outward sedimentary layers eroded to curved ridges and valleys. **Synclines** are downfolds or downarched folds. Take the sheets off the table and cause a buckling downward: There you have a syncline. Inclined beds here dip toward one another. Downfold counterparts of domes are **basins.** Keep in mind these are structural, not topographic, basins; that is, such basins result from the down-bending of rocks, not simply from the lowering of elevation. Basins and domes vary from less than a mile to several hundred miles in extent.

Many larger folds result from horizontal squeezing or compressive stresses occurring at considerable depth, where high confining pressure and temperature reduce the strength of rocks. Smaller domes, however, can be produced by other means, as by igneous or salt intrusion. Intrusion of molten rock or magma can form laccoliths (see Chapter 15) that dome up overlying strata. Intrusion of salt is more difficult to perceive, but does exist. Again, considerable pressure—in this case downward pressure from the

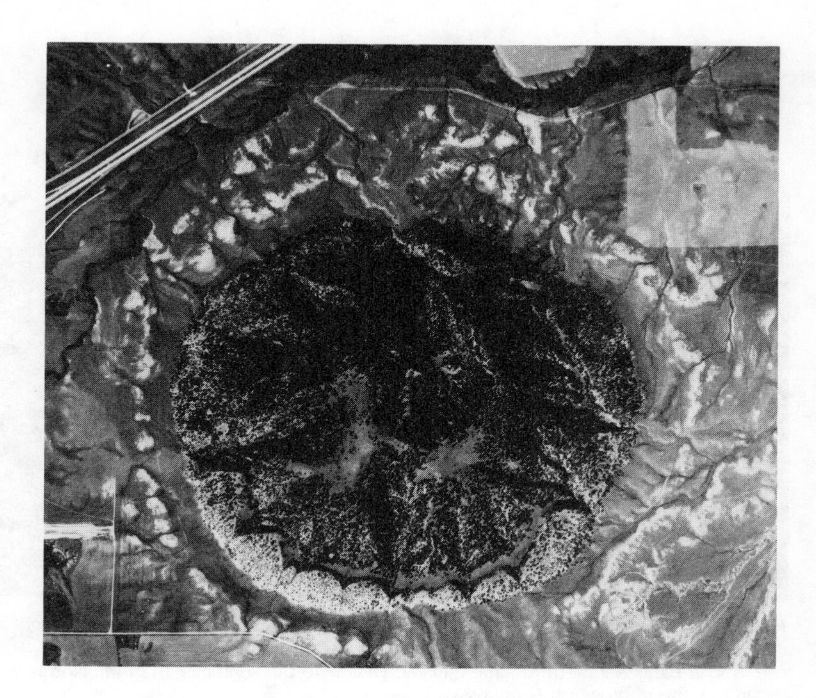

FIGURE 9–2. Vertical aerial (*upper*) and ground views of a nearly circular dome, Green Mountain, near Sundance, northeastern Wyoming. The darker pine-covered central part consists of sandstone, dolostone, and shale. Lighter, noticeably tilted beds with flatirons that flank the uplift, with fewer trees, are of limestone. The outermost, lightest beds lacking trees are of shale and rock gypsum; a butte of these rocks projects at the right in the lower photograph. The dome was punched up by a buried, intrusive laccolith (see Chapter 15). Width of the tree-surfaced dome is 1.2 miles (1.9 km). (Upper photograph is 278-146 of the U. S. Department of Agriculture, ASCS.)

weight of overlying strata will do—punches up pillars or columns of less dense salt from a thick "mother" layer. Rock layers overlying the salt pillars are domed upward. Hundreds of these domes are known to exist within the United States Gulf Coast area, both on land and submerged. Two visible at the surface—circular and about 2 miles (3.2 km) in diameter—are the Avery and Weeks Islands salt domes in southern Louisiana.

Now, let's picture downward erosion through anticlines and synclines, which usually occur in groups. Result: parallel ridges held up by resistant rocks with intervening valleys on weaker rocks (see Figure 9–1). Ridges developed from relatively steeply dipping

FIGURE 9–3. Hogback, Mt. Rundle, held up by limestones dipping steeply to the right and cut through by the Bow River. Flatirons are evident in the left distance. In the near foreground, smaller Tunnel Mountain rises above Banff, southern Alberta, Canada; view is to the southeast from Mt. Norquay. (Source of the photograph is unknown.)

beds form **hogbacks;** where cut through by regularly spaced streams or gullies, triangular segments called **flatirons** (Figure 9–3)—resembling the old style of iron for pressing clothes—result. Ridges of lesser-dipping beds, presenting more surface area, are **cuestas** (Figure 9–4). On either type of ridge, clearly asymmetrical in cross profile, the gentler slope is on the side toward the direction of dip. This helps you distinguish anticlines from synclines on aerial photographs and topographic maps, or while flying. So, if asymmetrical ridges on the flanks of a fold have their gentler slopes directed away from one another, the fold is an anticline. The opposite is true for a syncline. Another approach is to look for V-shaped notches (see Figure 9–1)—seen looking down on an eroded surface—in the hogbacks or cuestas where streams or gullies cut through. Points of Vs point in the direction of dip;

FIGURE 9–4. Vegetated cuesta formed by resistant beds dipping gently to the left.

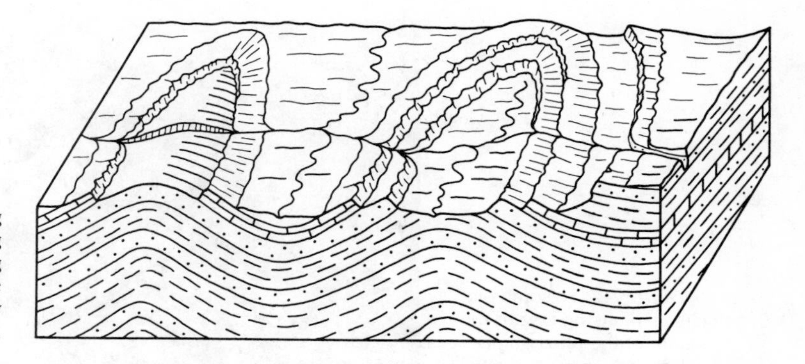

FIGURE 9–5. Eroded plunging anticlines and synclines, showing curved, converging and diverging ridges and valleys. Imagine the folds in Figure 9–1 tilted away from you and eroded.

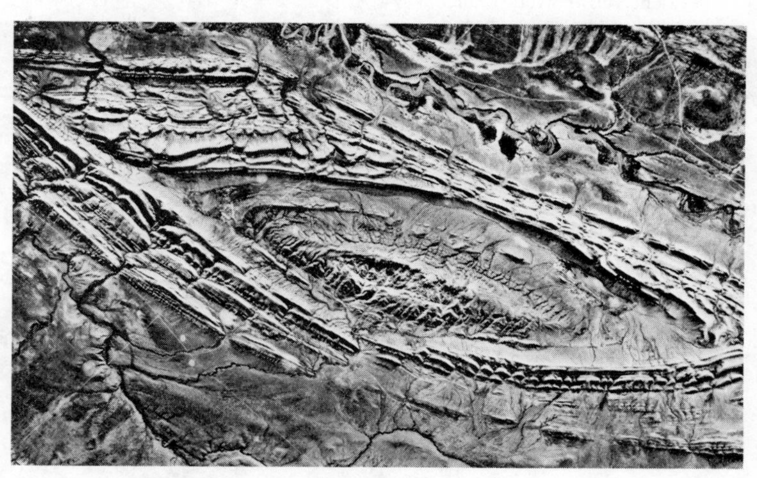

FIGURE 9–6. Eroded anticline plunging in two directions (to left and right), evidenced by elliptical hogbacks—with conspicuous V-shaped notches—and intervening valleys. Progressively younger shale and sandstone beds are found away from the center line or axis of the anticline. Maverick Spring Draw meanders in the upper part of the photograph. Little Dome, Fremont County, west-central Wyoming. (U. S. Geological Survey photograph Wyoming 7 A.)

therefore, Vs directed outward mean anticlines, those directed inward mean synclines. Long, narrow Vs indicate gentle dips; short, wide Vs signify steep dips. If no Vs exist, the beds are vertical.

Back to the sheets of paper. Off the table, create at least three folds. Now, tilt the folds along their lengths toward or away from you; this angle of tilt is the **plunge** of the folds. The next step may be hard to visualize at first: What do plunging folds look like on surfaces of erosion? In a nutshell, they produce a curved pattern of converging and diverging ridges and associated valleys (Figures 9–5 to 9–8); the pattern is zigzaglike where several anticlines and synclines occur together. If this is unclear, partly submerge your plunging paper folds in a sink or lake, whose surface simulates one of erosion. You will see a curved, zigzag line where the water's surface meets the paper. In the direction of plunge,

FIGURE 9–7. Folsom Point syncline—plunging toward the bottom of the photograph—evidenced by curved parallel ridges and valleys and V-shaped notches—where cut by streams. Northern Alaska. Progressively younger beds are found toward the center line or axis—drawn on the photograph—of the syncline. Utukok River cuts through the syncline in the foreground. (Photograph 366 by R. M. Chapman, U. S. Geological Survey.)

FIGURE 9–8. Shaded relief map of mountains and valleys resulting from the erosion of plunging folds. V-shaped notches along Green Mountain—to the right of Three Top Mountain—and asymmetrical slopes of the long ridge on the right of the map indicate a syncline underlies the valley between. This valley is about 2 miles (3.2 km) wide in the upper right. A syncline also underlies Little Fort Valley with the crest of an anticline along Green Mountain. All folds plunge to the lower left. Both the North and South Forks of the Shenandoah River—flanking the folds—display incised meanders. Near Strasburg, northern Virginia. (From the U. S. Geological Survey Strasburg Quadrangle, 1947.)

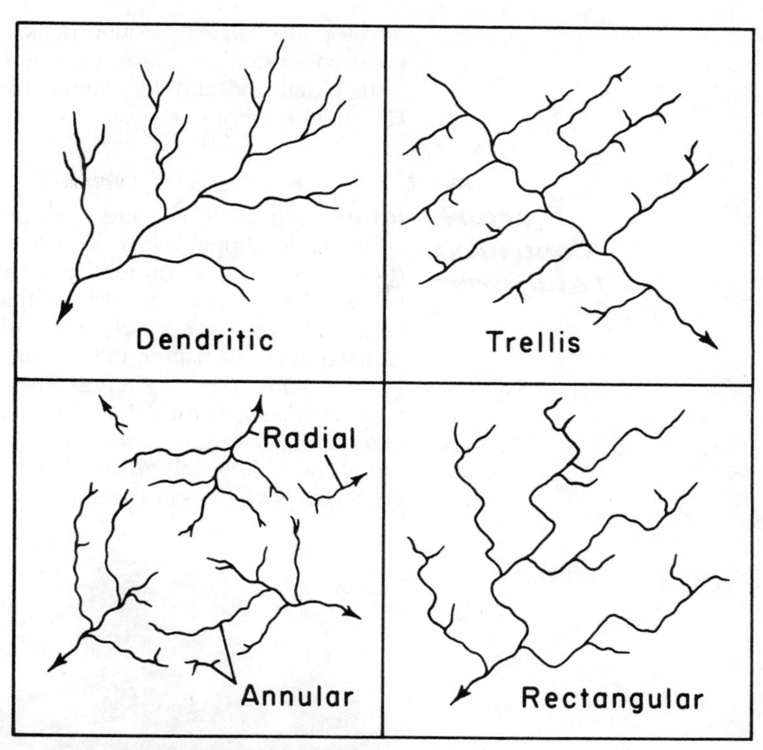

FIGURE 9–9. Main types of stream drainage patterns.

rock ridges converge toward the closed end of a loop in anticlines and diverge away from the closed end in synclines.

A characteristic **trellis drainage pattern** (Figure 9–9) of streams develops on associated, eroded anticlines and synclines—plunging or not. Similar in configuration to a trellis for climbing vines, long, parallel tributaries in valleys of weak rocks join the fold-transecting main streams at nearly right angles. The Ridge and Valley province of the Appalachian Mountains excellently portrays zigzag ridges and valleys on eroded plunging folds and a concomitant trellis drainage pattern. This is in marked contrast to the common **dendritic drainage pattern**—like irregular branches of a tree—that develops on flat-lying rocks or on uniform igneous or metamorphic rocks.

Center lines or axes of folds may be topographically higher or lower than the flanks, depending on the location of the most resistant rocks. Structurally high anticlines may have topographic lows at their crests, and structurally low synclines may be adorned with topographic highs (see Figures 9–1 and 9–5).

Domes and basins, as they erode, develop curved, alternate ridges and valleys in concentric or near-elliptical patterns. Initially, a **radial drainage pattern** (Figure 9–9)—streams diverge from the center—develops on a dome. (Such a drainage pattern may develop also on a mountain peak.) Later, an **annular drainage pattern**—ringlike—partly or largely replaces the radial pattern. Both radial and annular drainage have developed on the Black Hills dome mentioned earlier.

Fracture-Controlled Landforms

Joints (Figure 9–10) are fractures in rocks along which no appreciable slippage has occurred. Usually occurring in sets or directions of two or more, they divide rock layers into distinctive blocks. We are concerned here mainly with joints produced by large-scale stresses on rocks, not with those resulting from cooling of magma (see Chapter 15), or, say, unloading by erosion.

In some places, such as Zion and Arches National Parks in southern Utah, vertical joints cut sandstone beds (Figure 9–11) cemented by calcite. Ground and surface water seeps into the joints and enlarges them as the calcite dissolves. Water, wind, and gravity remove the sand grains. In time, rock slabs or fins—similar

FIGURE 9–10. Joints in the igneous rock felsite (see Chapter 15).

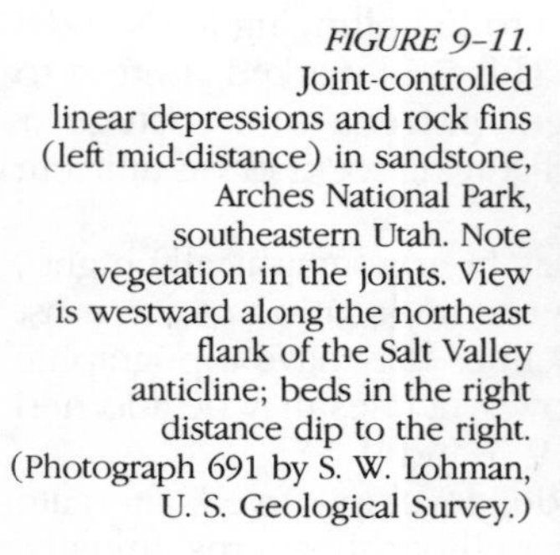

FIGURE 9–11. Joint-controlled linear depressions and rock fins (left mid-distance) in sandstone, Arches National Park, southeastern Utah. Note vegetation in the joints. View is westward along the northeast flank of the Salt Valley anticline; beds in the right distance dip to the right. (Photograph 691 by S. W. Lohman, U. S. Geological Survey.)

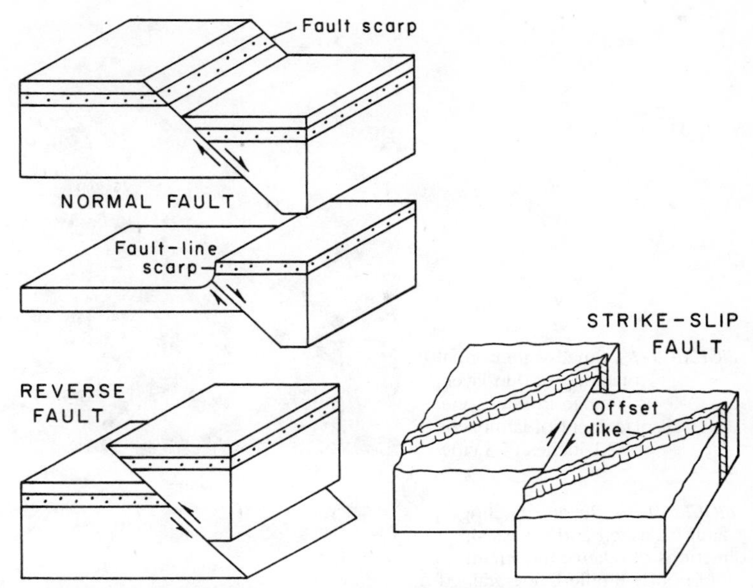

FIGURE 9–12. Main types of faults and associated features. In the lower normal fault diagram, a fault-line scarp owes its relief largely to later differential erosion. A resistant sandstone bed is dotted.

in shape to the back fins of fishes—are formed. Continued erosion by the agents just mentioned attacks the slabs or fins from the sides, forming alcoves that are ultimately cut through to produce natural arches (see Figure 13–1). These resemble the natural bridges in sandstone and limestone (see Chapter 6) that originate somewhat differently. Where intersecting joints are closely spaced, weathering (see Chapter 23) attacks from several sides to form rock pillars or columns. This phenomenon is particularly well displayed at Bryce Canyon National Park, also in southern Utah.

Joints may also control stream drainage, resulting in aligned, parallel valleys, or a **rectangular drainage pattern** (Figure 9–9) in which both main streams and tributaries make right-angled bends. Such a pattern develops in uniform rocks of any type as well as in foliated metamorphic rocks (see Chapter 17).

Faults (Figure 9–12) differ from joints in that appreciable slippage of rock masses has occurred along them. For faults that dip at a less than vertical angle, two main types are recognized. In a **normal fault** (Figure 9–13) the block above the fault has moved down relative to that below it. (I say "relative to" because the upper block may have moved down, the lower block may have moved up, or both may have moved in opposite directions.) In a **reverse fault** (Figure 9–14), the upper block has moved up relative to the lower block. A low angle reverse fault—say with a dip of less than 45°—is the common **thrust fault.** Normal faults are caused by pulling apart or tensional stresses, reverse faults by pushing together or compressive stresses.

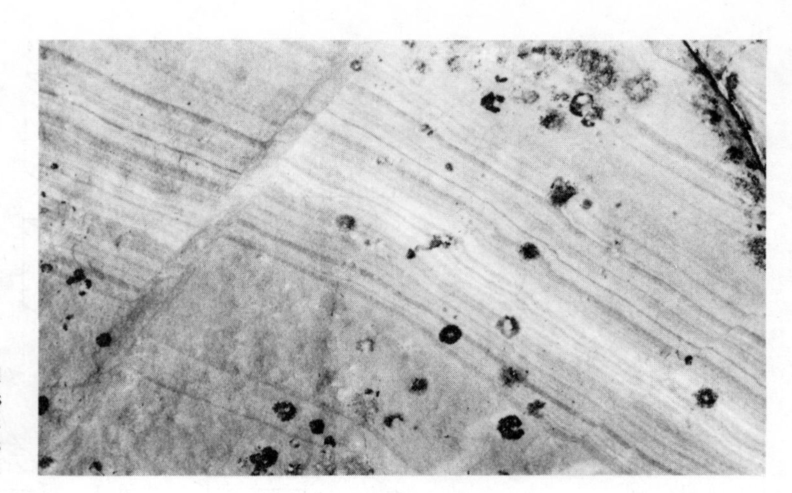

FIGURE 9–13. Small-scale normal faulting of thin layers or laminae in sandstone. Displacement of laminae is 2.1 inches (5.3 cm).

FIGURE 9–14. Reverse faulting—fault is dashed and arrows show directions of relative movement—of poorly lithified, fine-grained sedimentary strata. Displacement is at least 12 feet (3.6 m).

FIGURE 9–15. Glacier-sculpted, east-facing fault scarp of the Teton Range projects 7000 feet (2135 m) above low-lying Jackson Hole. Lacking foothills, the Tetons represent a block that moved up along a normal fault relative to the downward-moving block of Jackson Hole. Stream terraces flank the Snake River in the mid distance. Grand Teton National Park, northwestern Wyoming.

High angle normal and reverse faults form **scarps** (Figure 9–12; see also Figure 8–2), steep slopes or cliffs parallel to the faults. **Fault scarps** are due directly to faulting. Older **fault-line scarps** owe their relief to faulting plus later differential erosion, particularly where a more resistant bed contributes to the formation of such scarps (Figure 9–12, see normal fault). Fault scarps are associated with recent geologic features, as where they pass across alluvial fans or abut against scarp-dammed streams. They are short-lived and eroded by streams, glaciers, and gravity-induced processes. The eastern fronts of the Sierra Nevada Mountains in eastern California and the Teton Mountains in northwestern Wyoming (Figure 9–15) are highly eroded fault scarps. A fresher scarp faces west on the western edge of the Wasatch Mountains in northern Utah. Many scarps, of course, are considerably smaller.

Two normal faults dipping toward one another produce a down-dropped block or **graben** resulting in a straight-sided *rift valley.* Frequently, normal faults occur in groups, forming a series of grabens with intervening high, linear plateaus or mountains. The Basin and Range province covering Nevada and environs is such a region of generally north-trending grabens and small

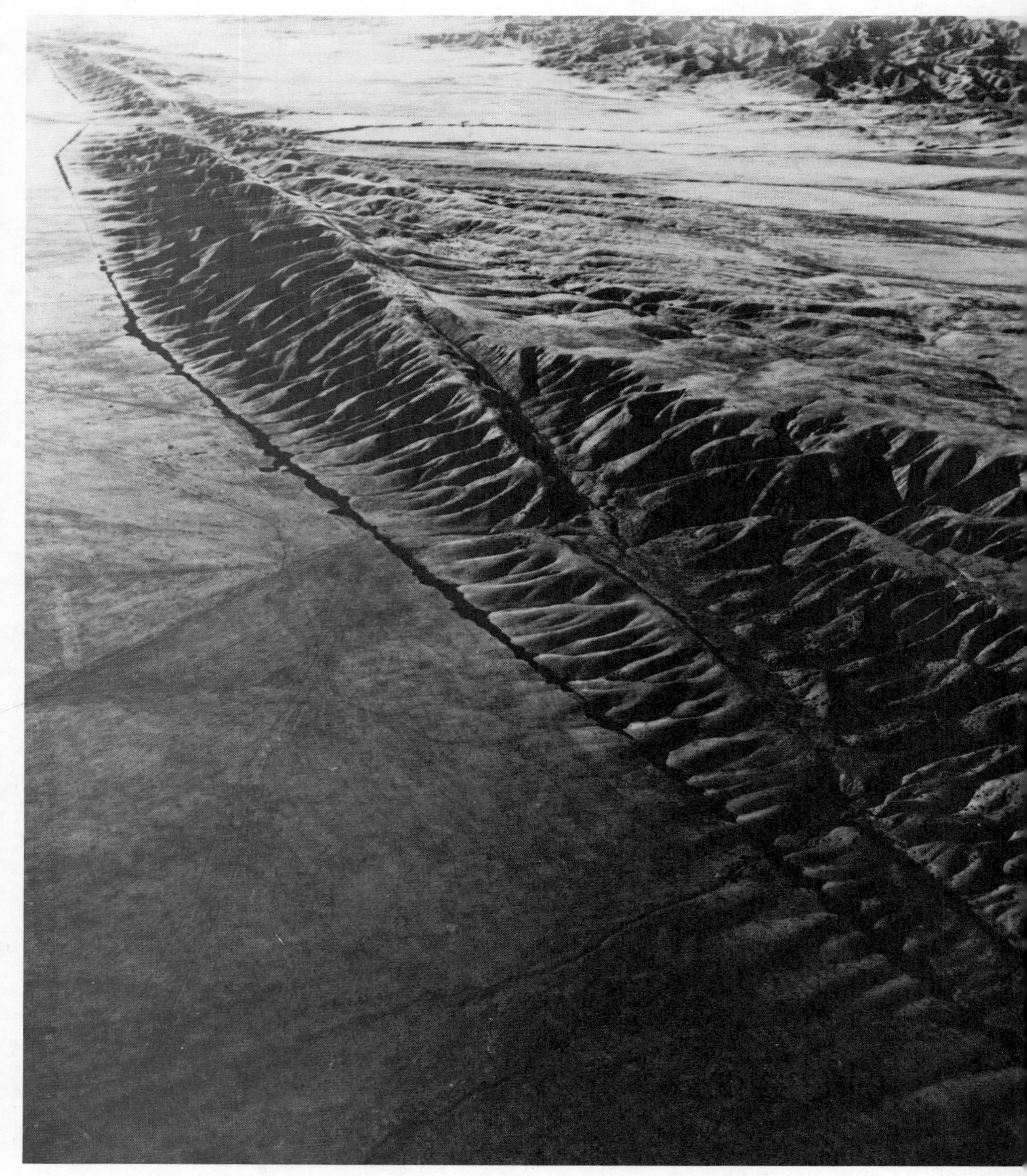

FIGURE 9–16. San Andreas strike-slip fault zone, San Luis Obispo County, southern California. (Photograph 194 by R. E. Wallace, U. S. Geological Survey.)

mountain ranges. In eastern Africa the rift valleys with lakes such as Tanganyika occupy grabens, as does the Dead Sea in Jordan.

In **strike-slip faults** (Figures 9–12, 9–16) displacement of rock masses is mostly horizontal and parallel to the strike or trend of the faults. A famous example is the San Andreas—actually a zone of faults—extending from the Gulf of California generally north-northwestward to north of San Francisco. Movement along strike-slip faults is evidenced by offset dikes, ridges, fences, streams, orchards, and other features. Movement to the right (Figure 9–12) is the case for the San Andreas. That is, if you stand on one side of the fault and look across to the other, linear features are seen shifted to the right. San Francisco rests on the east side of the San Andreas, Los Angeles on the west. A wag has said that in the geologic future the two cities, carried along on their respective blocks or earth plates, will be one!

Because of little or no up or down movement, landforms of strike-slip faults are inconspicuous. Shallow, linear valleys and low, linear ridges—from slivers of rock caught along a fault—are the main features. Offset ridges and stream valleys are associated landforms.

Like joints, faults can produce linear valleys, single or in parallel groups. Fault systems may also intersect so as to control a rectangular drainage pattern.

10 KEYS FOR IDENTIFICATION OF LANDFORMS

This chapter is a concise grouping of the main landforms, those covered in Chapters 2 through 9 but others as well, particularly those produced by man. Mainly, though, it is an attempt to provide quick identification of landforms through the use of keys. Biologists, particularly, use keys as solving devices to identify plants and animals.

These keys are based primarily on the setting of a landform, followed by composition and other characteristics. Each numbered step in a key offers you two choices. If the first choice doesn't provide the answer, a second choice—by means of a number—directs you to another couplet of choices. If you seem to end up with an illogical answer, backtrack through the key to where you most likely selected a wrong choice.

Identifying landforms produced by a geologic agent no longer present is of some difficulty. Examples of these would be shoreline landforms of a lake evidenced now only by a lake plain (its former bottom); or glacial landforms produced by a glacier melted long ago. Before using any of the keys—should you skip to them directly—it might be a good idea to at least skim Chapters 2 through 9.

Landforms in the keys are grouped as to whether they project upward (hills, ridges, and mountains); are relatively flat (plains and related features); or form depressions (valleys and basins). *Hills* are about as long as wide, *ridges* notably elongate; *mountains*, of either shape, are considerably larger than hills and ridges and reach heights of about 1000 feet (300 m) or more. *Plains* are relatively flat areas, horizontal or gently sloping, and frequently low-lying. Hills, ridges, or depressions may be present but the relative flatness dominates. *Valleys* are notably elongate depressions, usually open at one or both ends. *Basins* are normally about as long as wide and enclosed.

HILLS

1. Restricted to base of slopes
   Landslide hill (Figure 7–8): Scarps or cliffs above; of rock or rock debris.
   Talus (Figure 7–4): Rock accumulation, commonly in half-cones.
   Not restricted to base of slopes — 2

2. Associated directly with stream drainages
   Plateau: Large, elevated, wide, tableland.
   Mesa: Tableland smaller than plateau.
   Butte (Figure 9–2): Small, isolated, erosional remnant, particularly in badlands areas.
   Monadnock (Figure 2–7): Isolated, erosional remnant on a low-relief peneplain.
   Not associated directly with stream drainages — 3

3. Restricted to shorelines
   Stack (Figure 4–3): Isolated, associated with sea cliffs, arches, caves, and hanging valleys; of rock.
   Not restricted to shorelines — 4

4. Involves domed rock layers
   Dome hill: Resistant rock layers punched up by salt or laccolith (Figure 15–18) intrusion or squeezing from sides.
   Does not involve domed rock layers — 5

5. In areas of hot springs or geysers
   Thermal mound or cone (Figure 6–7): Of travertine or siliceous sinter.
   Not in areas of hot springs or geysers — 6

6. Bordered by terraces or flat surfaces
   Terraced hill: In rock from mining and in sediment and soil from farming.
   Not bordered by terraces — 7

7. Of rock
   Exfoliation dome (Figure 3–6): Slabs or shells weather and spall off massive rock to form rounded hill.
   Rock knob: Resistant rock of any type forms hills not covered previously.
   Pyramids, temples: Man-made.
   Not of rock — 8

8. Of till (nonuniform mixture of boulders through clay) or man-made fill
   Till knob: In moraines (Figure 3–11); of till; with foreign boulders or erratics.
   Mine spoil pile: Of fill.
   Burial or construction mound: Of fill.
   Of sand and gravel
   Kame (Figure 3–15): Steep-sided; formed in contact with glacial ice.
   Dune (Figure 5–8): Inland or near coasts; of sand only.

RIDGES

1. Restricted to base of slopes
   Landslide ridge (Figure 7–9): Scarps or cliffs above; of rock or rock debris.
   Not restricted to base of slopes — 2
2. Associated directly with streams
   Divide between streams (Figure 2–7): Of rock or sediment.
   Natural bridge: Usually of rock; formed by surface and underground streams.
   Natural levee: Flanks stream; of silt and sand.
   Spoil bank: From stream dredging or stream mining; may appear segmented.
   Not associated directly with streams — 3
3. Restricted to shorelines — 4
   Not restricted to shorelines — 5
4. Of rock
   Sea arch (Figure 4–2): Associated with sea caves and cliffs and hanging valleys.
   Of sand or gravel
   Beach (Figure 4–4): At shoreline and parallels it.
   Spit (Figure 4–4): Projects into open water.
   Baymouth bar (Figure 4–4): Blocks mouth of bay.
   Tombolo (Figure 4–4): Connects island to mainland or other island.
   Barrier island (Figure 4–4): Separated from low coast by lagoon.
5. Restricted to mountains
   Arête (Figures 3–1, 3–5): Sharp-crested, between half-bowl–like depressions (cirques).
   Not restricted to mountains — 6
6. On lava flows
   Pressure ridge
   Not on lava flows — 7
7. Involves tilted rock layers
   Hogback (Figure 9–3). Rock layers steeply inclined.
   Cuesta (Figure 9–4). Rock layers gently inclined.
   Does not involve tilted rock layers — 8
8. Of rock
   Sheep rock (Figure 3–9); Streamlined; surface scratched or grooved.
   Stone wall: Man-made, as in New England area.
   Dike wall (Figure 9–12): Of resistant rock of igneous dike.
   Of sediment — 9
9. Of till or man-made fill
   Drumlin (Figure 3–9, 3–13): Streamlined; of till.
   Moraine (Figures 3–8, 3–10): Lateral, medial, end, recessional; may be associated with glaciers; of till with foreign boulders or erratics.
   Road or railroad embankment: Where abandoned, look for straight sides; human rubbish or railroad ties or rails may be nearby; of fill.
   Mine spoil pile: Especially in coal-mining areas; of fill.

Of sand or gravel

Esker (Figure 3–14): Twisting; formed in contact with glacial ice.

Crevasse filling: Relatively straight; formed in contact with glacial ice.

Dune (Figure 5–1): Inland or near coasts; of sand only.

MOUNTAINS

Mountains typically occur in chains or ranges but also as isolated peaks. Peaks become jagged from secondary erosion by glaciers.

1. Entirely of volcanic rocks

   Volcanic mountain (Figure 8–4): Commonly as separated peak.

   Not entirely of volcanic rocks 2

2. Involves domed rock layers

   Dome mountain (Figure 9–2): Resistant rock layers punched up by laccolith intrusion or squeezing from sides.

   Does not involve domed rock layers 3

3. Interior of conspicuously folded and faulted rock layers

   Folded mountain (Figure 9–8): In long belt, most extensive of mountains.

   Interior not of conspicuously folded and faulted rock layers 4

4. Straight and linear with linear depression on one or both sides

   Fault-block mountain (Figure 9–15): Faulting along one or both sides of mountain, adjoining depression from down-dropped block.

   Not straight and linear with linear depression on one or both sides 5

5. Rock layers essentially flat-lying

   Erosional mountain: Result primarily of differential erosion with accompanying uplift; may be plateau, mesa, or monadnock (see *Hills*).

PLAINS AND RELATED FEATURES

1. Associated directly with streams

   Flood plain (Figure 2–2): Flat part of broad valleys inundated during floods.

   Stream terrace (Figures 2–4, 9–15): Benchlike remnant of former flood plain flanking valley walls.

   Delta plain (Figure 3–10): At mouth of present or former stream, slopes toward sea or lake; surface of delta.

   Alluvial plain (Figures 2–4, 2–6): Surface of alluvial fan or grouped fans; also, any plain of alluvium or stream sediment.

   Outwash plain (Figure 3–10): Near irregular knob and basin topography of glacial moraines; of melt water stream sediment.

   Valley train plain: Outwash plain confined to valley; present stream may be underfit for valley.

   Not associated directly with streams 2

2. Allied with shorelines

   Wave-cut bench (Figure 4–2): Slopes away from sea or lake cliff; forms terrace where raised above water level; of rock.

   Wave-built terrace (Figure 4–2): Seaward or lakeward of

wave-cut bench; not readily visible except of former water body; of sediment.

Tidal flat: Periodically covered and uncovered by rising and falling tide.

Coastal plain: Slopes toward the sea; generally represents raised former sea bottom.

Lake plain (Figures 3–17, 3–18, 3–10): Associated with shoreline features, former lake bottom; fertile; of silt and clay; commonly direct or indirect result of glacier formation and extension.

Not allied with shorelines 3

3. Restricted to lava flows

   Lava field or plain (Figure 8–2): Broad expanse underlain by lava flows; may be associated with volcanic cones.

   Not restricted to lava flows 4

4. Of till

   Till plain: Of low hills and shallow, closed depressions; corresponds to ground moraine (Figures 3–10, 3–12).

   Not of till

   Pediment: Planed rock surface sloping away from mountain front or cliff; may be veneered with stream sediment; in dry regions.

   Peneplain (Figure 2–7): Planed rock surface unrelated to mountain front or cliff, often with scattered erosional remnants (monadnocks).

VALLEYS

1. Restricted to soluble rocks

   Solution valley (Figure 6–2): Blind or closed; most common in limestone regions; caves nearby.

   Not restricted to soluble rocks 2

2. Straight-sided and usually flat-bottomed

   Rift valley: Bounded by steep cliffs or linear mountains; result of faulting, valley occupies down-dropped block.

   Not straight-sided and usually flat-bottomed 3

3. Tributary valleys present

   Stream valley (Figure 2–7): Tributary valleys meet main valley at its level; narrow valley with V-shaped cross-valley profile; may be straight where it follows fracture or tilted rock layers (Figures 9–9, 9–16); may lack stream or one that is underfit as in abandoned melt water valley.

   Glacial valley (Figure 3–6): Tributary valleys meet main valley above its level; cross-valley profile U-shaped, bottom may be flattened from sediment fill; may be secondarily occupied by stream; *fiord* is drowned glacial valley.

   Tributary valleys generally lacking

   Ditch, canal, channel: Sides even; may be lined with rock, concrete, or other material; for irrigation, drainage, diversion.

BASINS

Most basins are wholly or partly filled with water to form lakes or ponds.

1. Restricted to stream valleys — 2
   Not restricted to stream valleys — 3

2. In rock
   Pothole: Usually deeper than wide, in stream channel at sites of present or former waterfalls and rapids; worn by sand and gravel in whirling water.
   In sediment
   Meander scar (Figure 2–2): Part of curved abandoned stream bend in broad valley, *oxbow lake* if filled with water.
   Backswamp: Shallow depression on flood plain away from stream in broad valley.

3. Restricted to folded or faulted rock layers
   Tectonic basin: From downfolding or downfaulting.
   Not restricted to folded or faulted rock layers — 4

4. Restricted to volcanoes
   Volcanic crater (Figure 8–1): At top of volcano.
   Caldera (Figure 8–2): Similar to volcanic crater but larger.
   Not restricted to volcanoes — 5

5. Overlies soluble rocks
   Sinkhole (Figure 6–3): Solution cavity at surface; most often in or above limestone; caves nearby.
   Does not necessarily overlie soluble rocks — 6

6. Restricted to glacial valleys (see *Valleys*)
   Cirque (Figures 3–5, 3–7): Half-bowl–like depression at head of glacial valley, commonly with distinct basin at base.
   Not restricted to glacial valleys — 7

7. Usually bordered by raised rim
   Meteor crater: Rock in crater may show effects of shock metamorphism (see Chapter 17); rare on earth, common on moon and other planets.
   Bomb crater: Areas of human conflict; associated with human rubbish and relicts of war.
   Not usually bordered by raised rim — 8

8. Restricted to desert floor
   Playa: Contains lake after rains, accumulates salts.
   Not restricted to desert floor — 9

9. Characteristic of areas of mining
   Quarry, open-pit mine, sand-gravel pit: Site of extraction of valuable rocks or minerals; associated with *spoil piles;* open-pit mine may be terraced.
   Collapse (subsidence) pit or basin (Chapter 7): Commonly associated with spoil piles; sinking may cause settling of buildings and local flooding; collapse pits related to underground mines elongate or aligned. May also form by overwithdrawal of ground water or oil.
   Not characteristic of areas of mining — 10

10. In rock

Rock basin (Figure 3–5): Grooves and scratches indicate formation by ice scour; scattered along glacial valleys and in other glaciated areas where rock is exposed.

In sediment

Blowout: In hilly area of sand; sides often steep, cut layering of adjacent dunes.

Kettle (Figures 3–10, 3–16): In hilly area of till and relatively flat area of sand and gravel; steep-sided, at least in part; result of melting of partly or wholly covered block of ice.

Swale (Figure 3–12): Similar to kettle in setting but not steep-sided, result of uneven laying down of sediment.

Watering pond basin: Man-made; with bordering spoil banks; livestock may be nearby.

Contemplating the past

11 TIME AND GEOLOGY

Time, the elusive substance. We "make time," "mark time," "kill time." But what really *is* it? I like this Webster's definition: "Every moment there has ever been or ever will be." No stipulation as to a beginning for time or an end. Most of us view time as irretrievable and irreversible, although physicists delve into its possible reversibility. But time by itself, to humans, is essentially meaningless; it gains significance when related to a happening or event. In this sense, time is a period or interval when something exists or happens, or the period or interval between a thing's nonexistence and existence, not happening or happening. We've come to accept the inseparability of time and events.

Time can be conceived in a relative or specific sense. We speak of a happening in the past, present, or future, or occurring before, at the same time as, or after another. That's relative. Specific time for an event is reckoned in years, months, days, hours, minutes, seconds.

We witness the passage of time with observations of birth and death, day and night, and the changing of the seasons. We are especially attuned to the earth's rotation and revolution. Landscape changes reflect time's passage, too, but more slowly: the cutting of stream valleys; the molding of the surface by advancing and retreating glaciers; and the shaping of shorelines, by erosion in places, by deposition elsewhere.

Historic time—the period of written history—goes back about 5000 years. Anything prior is prehistoric time. In the seventeenth century, Biblical scholars stipulated that prehistoric time extended back to about 6000 years; this figure prevailed until the nineteenth century when several dating attempts, based on scientific principles, pushed the beginning of prehistoric time back to millions of years earlier. Geologists have since demonstrated that *billions,* not thousands or millions, of years encompass prehistory. Prehistoric time since the earth's origin is commonly referred to as *geologic time.* It goes back to when the earth attained most of its mass—and perhaps a solid crust—by, presumably, condensation of a whirling, eddying cloud of dust and gas, the process that formed other components of the solar system. The universe, of which the solar system is but an infinitesimal part, may have originated billions of years earlier. Time since the formation of this all-encompassing expanse may be called cosmic or universe time.

Measurement of Time

To measure and record time, we are not restricted to human clocks and calendars, and, in fact, we cannot be for prehistoric time. Well-known, natural time recorders reflecting seasonal changes are tree rings and "growth" or rest rings on clam shells. The bristlecone pine of the American Southwest and eastern California is a particularly ancient tree. Counting rings of dead and living trees has extended a record of time well beyond the 6000 years originally believed as the age of the earth. **Varves,** or yearly beds, also reflect annual seasonal changes and, therefore, record time. The best-known varves are those laid down in lakes formed from glacial melt water and damming. Each varve is of a lighter, thicker, coarser summer layer and a darker, thinner, finer layer deposited during ice-covered periods. You simply count the couplets of beds for the span of a lake's existence. Varve counting of glacial lake sediments relating to last glaciations gives values to about 15,000 years ago; that of certain older, nonglacial lakes gives their duration as a few million years. But all the methods we've looked at so far are restricted in the magnitude of time they measure or their general applicability. **Radioactive** or **radiometric dating,** however, with its several methods, allows us to measure virtually any interval of time involving earth materials and events.

At about the turn of the twentieth century, certain elements or varieties of elements—or **isotopes** that differ from their counterparts by having different atomic weights—were found to be unstable or *radioactive.* That is, they decay spontaneously as charged particles are emitted from the centers or nuclei of atoms. In so doing, one element transforms into another. The rate of decay seems to be constant with time and essentially unaffected by changes in temperature, pressure, or chemical changes. This rate for a radioactive element is expressed as *half-life,* the time in years for half of the original number of atoms to decay. Uranium 235 (the number refers to the atomic weight), one of the isotopes of uranium, has a half-life of 713 million years. After this time, half of the atoms of this isotope in a sample are gone; in two half-lives or 1426 million years, half of a half or only a quarter remain.

To date a mineral or rock with a radioactive element, we must know the relative amount of the derived or daughter element in relation to the amount of the original or parent element. Sophisticated equipment can measure this. Uranium 235 ultimately decays to lead 207. If a mineral like zircon in an igneous rock contains 50 percent derived lead and 50 percent original uranium, one half-life has gone by, and the dated sample is 713 million years old. Seventy-five percent derived lead and 25 percent uranium indicate two half-lives have elapsed for a date of 1426 million years; and 87.5 percent lead and 12.5 percent uranium tells us three half-lives have elapsed or the sample has an age of 2139 million years.

The chief radioactive alterations used in dating minerals and rocks are: uranium to lead, potassium to argon, rubidium to strontium, thorium to lead, and carbon to nitrogen. All methods except carbon to nitrogen can date the oldest rocks because of long half-lives; the longest involves rubidium 87 with a half-life of 47,000 million years. Carbon to nitrogen or the carbon 14 method operates somewhat differently and can date only organic materials (those that contain carbon, such as wood, bone, and shell). Carbon 14, formed as nitrogen is bombarded by cosmic rays in the upper atmosphere, is absorbed by every living organism after it combines with oxygen to form carbon dioxide. When an organism dies, carbon 14 is no longer replenished and begins to diminish at its decay rate, according to a half-life of about 5730 years. A date is based on the ratio of carbon 14 to all other carbon in a sample; the less carbon 14 found, the greater is the age. Because of the short half-life involved, the carbon 14 method can date organic materials only less than about 100,000 years old.

Radiometric dating is a neat, useful technique, but I would be remiss if I didn't point out its inherent fallibility. That doesn't mean we shouldn't believe the derived dates; we must simply apply them cautiously. A major source of possible error involves the assumption that an analyzed sample has persisted within a closed system; that neither original nor derived materials were removed or added except by radioactive decay. This implies, too, that no derived material was originally present. If this were not the case, the calculated date, of course, would be incorrect. This possible error is corrected for by using different dating methods, checking one against another. Then there is error from laboratory analysis. Every date is reported with a plus or minus figure, such as 430 million years ± 10 million years. Plus or minus 10 million years may be strictly from error in measurement or reflect variations in several possible samples analyzed. It means another analyzed sample would probably fall within 10 million years on either side of 430 million, but says nothing about how good the radiometric age is in relation to the true age. The true age is approached, again, by further analyses by different methods. At first, ± 10 million years may seem a large laboratory error. But it's less than five percent of 430 million years.

Geologic Time Scale

The geologic time scale (Figure 11–1, Table 11–1) was developed during the eighteenth and nineteenth centuries, mostly in Europe where the science of geology was born. Development took place rather haphazardly, with no preconceived plan. Names of the time subdivisions were inconsistently derived after places, a mountain range, ancient tribes, or rocks. Initially, vertical sequences of sedimentary strata in selected places revealed their relative age through **superposition:** In an undisturbed sequence the oldest rocks are at the bottom and the youngest are at the top. But how to piece together many rock sequences to arrive at a relatively complete, unified whole? (No place on earth preserves a complete rock record of geologic time.) Here, the principle of **faunal succession** came to the rescue. This assumes that fossil assemblages occur in a definite order or succession in a vertical sequence of rocks. This allowed the matching up or demon-

FIGURE 11–1. Geologic time scale. The scale on the right margin compares geologic time to days in a year. Selected biological events show that most occurred relatively recently. Subdivision of the Cenozoic is given in Table 11–1.

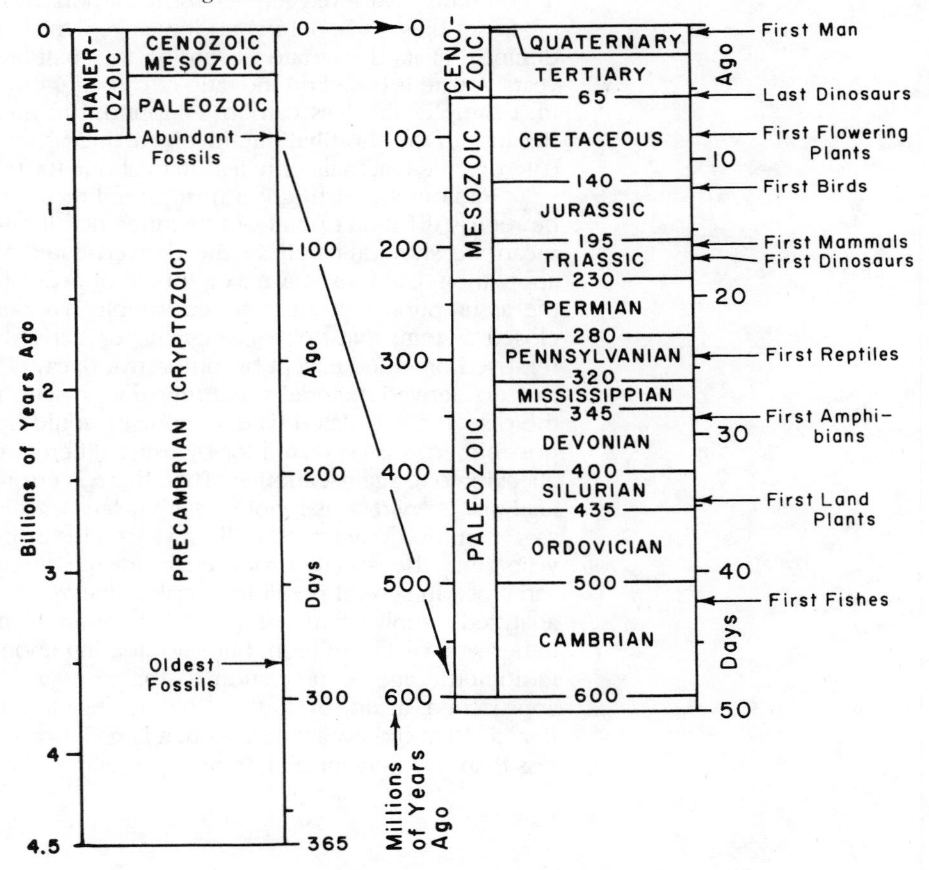

Table 11-1. SUBDIVISIONS OF THE CENOZOIC ERA WITH SELECTED BIOLOGICAL EVENTS

PERIOD	*EPOCH*	*YEARS SINCE BEGINNING*	*SELECTED BIOLOGICAL EVENTS DURING EPOCHS*
Quaternary	Holocene	10 Thousand	
	Pleistocene	2 Million	First Modern Man
Tertiary	Pliocene	5 Million	First Man
	Miocene	24 Million	First Abundant Grasses
	Oligocene	37 Million	
	Eocene	54 Million	First Rhinoceroses
	Paleocene	65 Million	First Horses

strating of equivalence—**correlation**—of rock layers in widely separated places. Realize at this point that rock layers were stacked up in their correct order—oldest to youngest—and subdivisions of the rock record eventually became subdivisions of the time record. After most of the relative time scale was formulated in Europe it was extrapolated to other parts of the world, by fossils, because rocks change from place to place.

But the geologic time scale needed calibration. This came about with radiometric dating of primarily igneous intrusives and volcanic rocks interlayered with sedimentary rocks. From the principle of **crosscutting relationships**—a rock or fault is younger than the rock it cuts—we can see in Figure 11-2 that the dike (5) is younger than layers 1, 2, and 3. If the date of dike rock is 120 million years ± 5 million years, layers 1, 2, and 3 are older than this value. The wavy line (4) is an **unconformity** or surface of erosion. Upon erosion, boulders of resistant dike rock were incorporated within layer 6, demonstrating the principle of **inclusion:** Fragments of older rocks are enclosed within younger rocks (see also Figure 15-14). A lava flow (7) encroached upon the region, later to be buried by layer 8. If the lava flow rock is dated at 15 million years ± 1 million years, layer 6 has an age somewhere between 120 and 15 million years. So, by a combination of relative dating principles and numerous radiometric dates, the rock record is continually scrutinized and the time scale becomes calibrated in more detail. The process still continues.

Now, let's take a close look at the time scale. All of geologic time is first subdivided into two **eons,** the Cryptozoic or Precambrian and the Phanerozoic. A most lopsided split, to be sure, for the Precambrian constitutes more than 85 percent of geologic time. Eons are subdivided into **eras**—Paleozoic, Mesozoic, and Cenozoic; eras, into **periods** (such as Cambrian); and periods, into **epochs** (such as Paleocene). Only epochs for the Cenozoic are given here because of disagreement over many of the others. Speaking of disagreement, much exists over the time boundaries of the subdivisions. Boundaries are tied to appearances or extinctions of organisms and such physical events as

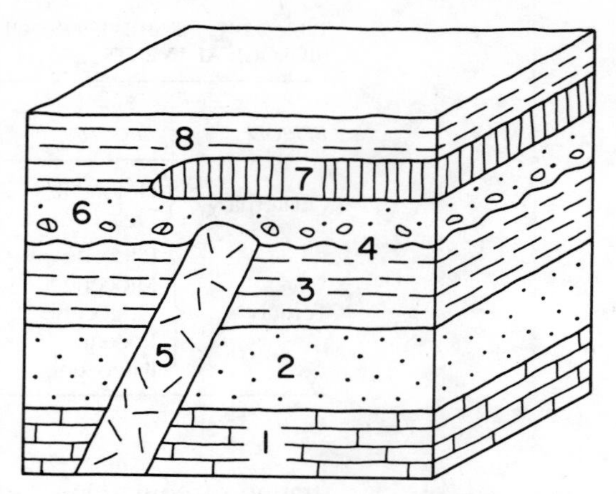

FIGURE 11–2. Diagram showing how relative and radiometric dating of rocks and events is used to formulate a time scale. Numbers indicate the correct order of events. The dike (*5*) (Chapter 15) is dated at 120 ± 5 million years, the lava flow (*7*) at 15 ± 1 million years. Layers *1* to *3* were laid down and intruded by the dike. Erosion, indicated by the wavy erosional surface or unconformity (*4*), occurred after 120 million years after which layer *6* was deposited, followed by emplacement of the lava flow. Layer *6* is older than 15 million years and layer *8*, deposited after the lava flow, is younger.

mountain building. Don't expect what you see here to necessarily agree with the data in other publications. But the disagreements involve only a few million years, which is no big thing when dealing with many millions or billions of years. This points out, too, that geology is a growing, steadily improving science.

Should you wish to learn the time scale, I might suggest using a jingle as an aid for the periods of the Paleozoic: *C*ome *O*ver *S*ome-*D*ay *M*aybe *P*lay *P*oker. (I learned this from my first geology professor, so it must be O.K.) If this seems helpful, create other jingles of your own for the rest of the time scale.

You might have wondered why the time scale begins 4.5 billion years ago—the approximate age of the earth. The oldest known rocks are not that old: 3.8-billion-year-old metamorphosed sedimentary rocks from southwestern Greenland, and perhaps some from western Australia that may be somewhat older. The oldest moon rocks, however, have been dated at about 4.6 billion years, as have been meteorites, believed by many to be remnants of a disrupted planet or fragments that never coalesced to form one. If the moon and meteorites are part of the same solar system as is the earth, the unity of all implies a common origin in process and presumably in time. So, the age of the moon and meteorites is the age of the earth.

Comprehending Geologic Time

No question about it. Comprehending the magnitude of geologic time is difficult. We simply lack a suitable frame of reference to relate to. However, although it is perplexing, we can try to understand it.

Many persons reach an age of a hundred, so let's consider that convenient even number as a beginning. If you lived ten times as long you would be a thousand years old. And, ten times more, to a ripe old age of ten thousand. Now, that's *really* old. But if you increased your age by increments of ten times *two* or *five more* times, your age would be one million or one billion years. Even at one billion, the earth is four and one-half times older.

Another way of attempting comprehension is equating the age of the earth to one calendar year (Figure 11-1) and considering when certain biological events occurred. The oldest fossils are known from a time corresponding to more than nine months ago, and life probably originated before then. But abundant life, or life capable of leaving a good fossil record, didn't exist until 48.5 days ago (beginning of the Cambrian). Dinosaurs lived and died out completely about 17 to 5 days ago, and recognizable humans appeared on earth within the last eight hours. Columbus barely discovered America this hypothetical year about three seconds ago.

Selected Readings

BERRY, W. B. N. *Growth of a Prehistoric Time Scale.* San Francisco: W. H. Freeman and Company Publishers, 1968.

EICHER, D. L. *Geologic Time.* Englewood Cliffs, N. J.: Prentice-Hall, Inc., 1971.

12 RECONSTRUCTING PAST EVENTS

All rocks are storytellers, providing the listeners are alert and attentive. Geologists are the listeners, and they spend much time attempting to unravel earth's history from the stories they hear or believe they hear. Heard incorrectly, a story may be largely fictitious, like a tested theory proven false. In time, though, a story's veracity may be confirmed by several listeners, each hearing the same tale or a bit more than the other.

Most geologists, reconstructing past events, accept the premise that processes controlling events today must have done so formerly, and events occurring today must have occurred in the past. (This concept is described in Chapter 19 as *uniformitarianism* or *actualism.*) But certain events of the past do not take place presently. The approach is useful and necessary, but must be applied with some care.

This chapter demonstrates how geologists attempt to decipher events from earth's varied history and provides raw material for your own contemplation of the past. Evidence for mountain building, former continental seas, and past climates is illustrated.

Mountain Building

We are concerned here with mountains that presently form linear belts, made of largely folded and thrust-faulted sedimentary rocks produced by squeezing and crustal shortening. Such mountain belts are not distributed haphazardly, but typically occur at or near the margins of continents (Figure 12–1). The Rockies and other mountain belts in western North America and the Andes in western South America are examples. A few, such as the Urals of Russia and the Himalayas of southern Asia, are found toward continental interiors. Mountain belts, too, are not all of the same age; the mountains of the western Americas and Himalayas formed during the Mesozoic and Cenozoic, but the Urals are late Paleozoic mountains. So, mountain building is a recurring event.

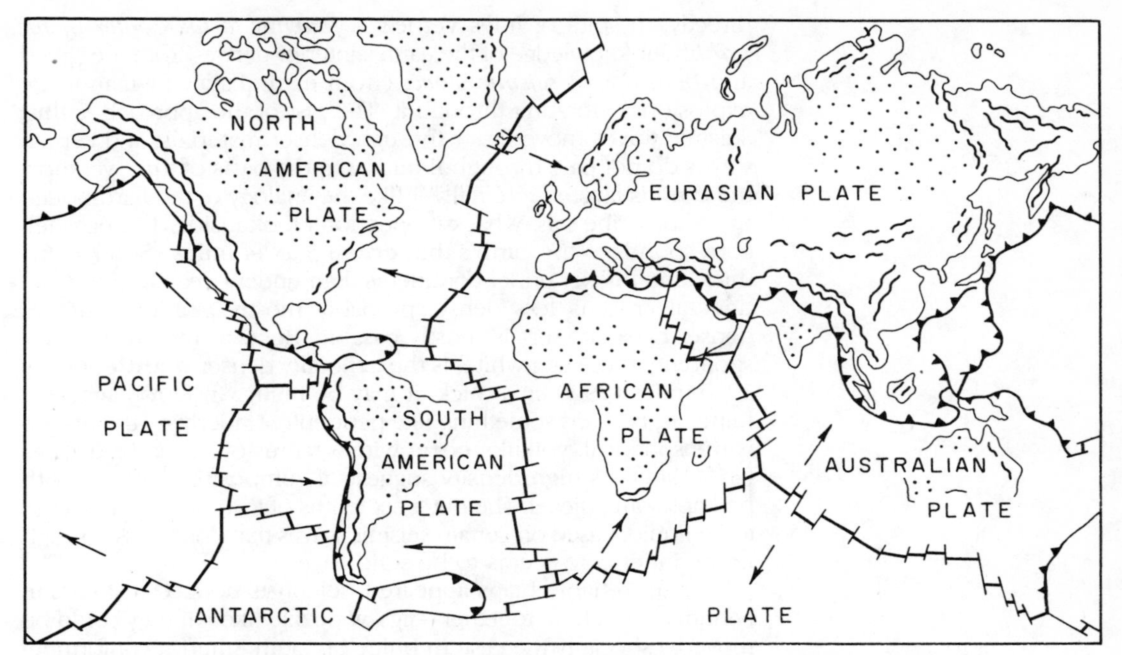

FIGURE 12–1. Tectonic map of the world showing lithospheric plates (seven major ones are named), shields (stippled), mountain belts (heavy, wiggly lines on continents: short lines indicate Paleozoic belts, longer lines locate younger belts), and stable platforms (blank areas on continents). The plates are bounded by the mid-oceanic ridge (straight, heavy line offset along fine lines that depict transform faults) and oceanic trenches or earlier subduction zones (heavy barbed lines). Arrows give the direction of plate movement.

While we're at it, let's look at the distribution of other large-scale crustal and deformational—or **tectonic**—features, and note their relationship to mountain belts. The nuclei of continents consist of **shields,** low-lying, low-relief areas of mostly Precambrian metamorphic and granitic rocks, jointed and faulted. In North America, the Canadian Shield is exposed over most of eastern Canada. Where not exposed, shields are covered by thin veneers of sedimentary rocks and are called **stable platforms.** Within platforms, rocks are flat-lying or warped regionally into broad, gentle basins and domes.

If you mentally drain away the ocean basins, the feature which perhaps first catches your eye is the mid-oceanic ridge, a mountain range in its own right: 40,000 miles (64,000 km) long and up to 9800 feet (3000 m) high. Unlike continental mountain belts, the ridge is made entirely of basalt, not of folded and faulted sedimentary rocks. A prominent rift valley occupies the crest along most of its length. The ridge is cut by numerous **transform faults,** a type of strike-slip fault. Generally near the margins of ocean basins, in places, are trenches reaching depths of nearly seven miles (11 km). Most conspicuous are those in the western Pacific from New Zealand to Japan.

To understand mountain building further, we must have a glimpse of the earth's interior. Obviously, we cannot gain this directly, in spite of Jules Verne's *A Journey to the Center of the Earth!* Our knowledge of the earth's interior derives from the speed and behavior of *seismic waves* (from earthquakes or man-made explosions) vibrating through it. The gist of the approach is this: Seismic waves move faster through denser materials, and certain waves do not pass through liquids. On the basis of this, we know the earth is layered internally. I like the analogy of the hard-boiled egg to describe this. Whack it open for a look inside. The eggshell corresponds to the earth's thin **crust:** 3 to 44 miles (5 to 70 km) thick, and thicker under continents than under ocean basins. Continental crust is less dense, probably mostly granitic, whereas denser oceanic crust is mostly basaltic. Beneath the crust, corresponding to the egg white, is the generally denser **mantle,** nearly 1800 miles (2900 km) thick; it may be higher in ferromagnesian minerals, perhaps something like peridotite. Underlying the mantle is the spherical, yolklike **core,** with a radius of about 2200 miles (3500 km); its high density suggests a composition of iron with perhaps some nickel. The outer six-tenths of the core is considered to be liquid, based on certain seismic waves that don't pass through it; the inner core seems to be solid.

After reliable maps appeared, scientists noticed that certain continents might fit together—jigsaw-puzzle-like—if they could be moved, especially the eastern bulge of South America conforming to the western reentrant of Africa. This observation and similar geologic evidence—such as fossils—on opposite sides of the Atlantic gave birth in the early 1900s to the theory of **continental drift:** Lighter continents drifted through heavier (denser) ocean-basin rock, possibly driven by forces resulting from the earth's rotation. In the 1960s, after the topography of the sea floor became known, drifting was replaced by the theory of **plate tectonics.** This posits that the earth's rind is divided into seven major, and many smaller, rigid plates (see Figure 12–1) jostling about, different parts of plates moving at different speeds. Each plate can be thought of as a fragment of our eggshell slipping on a sphere. But plates—about 60 miles (100 km) or so thick—involve the crust *and* uppermost mantle, or the **lithosphere.** Plates include both continental and oceanic crust—usually—or only oceanic crust (e.g., Pacific plate). They slide along on a weaker, mobile, perhaps partly molten **asthenosphere** (Greek *astheneia,* "weakness") within the upper mantle. You can analogize plates with sliding, crunching, slamming ice slabs upon ice breakup on a stream in spring. The stream current corresponds to a moving asthenosphere. What drives lithospheric plates? Most geologists seem to favor drag by convection currents brought about by heating within the asthenosphere or entire mantle. Heat may derive from decay of radioactive minerals within the mantle. Another possibility is heat radiated from the hot, outer liquid core. Visualize convection currents as those vertically circulating cells in a pot of simmering soup or those in cumulus clouds as the sun's rays heat a land surface. How does plate tectonics differ from drift? In plate tectonics, continents do not plow through ocean basins but ride along like ships frozen in floating ice sheets.

Considerable evidence shores up the plate tectonics theory, but we can examine only a bit of it here. One line of reasoning revolves about sea-floor spreading. Away from the mid-oceanic ridge, the linear region of basalt effusion with a crested rift valley, the sea floor spreads to either side. Deep-sea drilling and seismic exploration show sediments are thickest along the ridge's flanks and thinner toward its crest. In addition, sediments directly overlying the basalt become progressively older away from the ridge on either side. Both thickness and age trends support sea-floor spreading away from the ridge. Iceland straddles the ridge, and here the rate of spreading is measured directly across the exposed rift valley. Measurements here and indirect measurements elsewhere show sea-floor spreading or plate movement as less than an inch to a few inches per year.

Another line of evidence for plate tectonics comes from earthquake distribution. Most earthquakes occur in narrow zones along the mid-oceanic ridge, oceanic trenches, and young mountain belts. Those along the mid-oceanic ridge and associated transform faults have shallow origins—less than about 44 miles (70 km) below the earth's surface; those along trenches and mountain belts extend much deeper and follow a zone inclined toward a continent or an arc of volcanic islands. Earthquake zones logically are assumed to delineate plate boundaries that also correspond to belts of major volcanic activity. You might expect all this activity where plates are continually bumping into one another.

Plates, then, move away from the oceanic ridge where oceanic crust is created and parallel to the transform faults that offset the ridge in many places. (Transform faults, therefore, indicate the direction of plate movement.) At trenches, plates dip down into the mantle; in these **subduction zones** oceanic crust goes down—because it is heavier or denser—and is melted or assimilated and destroyed. Along transform faults, passively shifting crust is neither created nor destroyed.

You may be wondering: When are we going to get on with mountain building? We have finally arrived, though our arrival may be nearly anticlimactic by this time. But it need not be. You may have caught a hint or two along the way. Remember that squeezed mountain belts—especially the younger ones—concentrate at or near the margins of continents and earthquakes concentrate along young mountain belts, especially the deep-origin ones, along with volcanism? So, from this, plate tectonics theory postulates that mountain building occurs where converging plates slam into one another. Let's examine a case where one plate with continental crust and another with oceanic crust converge (Figure 12–2).

A thick sequence—thousands of feet—of sediments develops along the seaward margin of the continental plate: shallow-water sediments on the continental shelf and slope and deep-water sed-

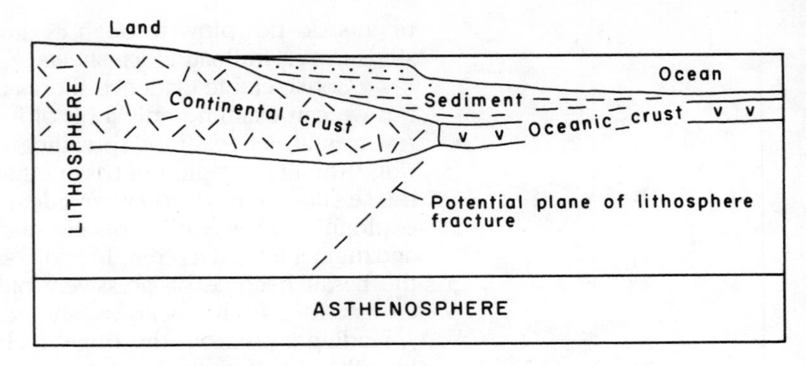

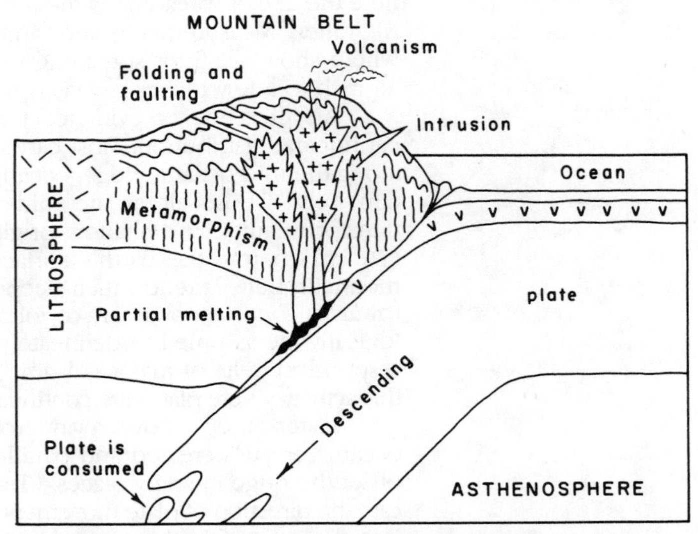

FIGURE 12-2. Mountain building by collision of plates separately including continental and oceanic crust. As the heavier oceanic plate descends, sediment (upper) and crustal rocks are squeezed into an elevated mountain belt that is folded and faulted, metamorphosed, intruded, and associated with possible volcanism.

iments on the continental rise, as well as thinner sediments on the deep ocean floor. As the two plates merge, the heavier oceanic plate is subducted. The sedimentary layers of both plates are squeezed and raised into a huge linear welt, thrust faulted at or near the surface where rocks are brittle, folded at a depth where flowage occurs. Part of the oceanic crust—made of deep-sea sediment, basalt, gabbro, and peridotite—is scraped off the descending plate and plastered against the compressed welt. This slice of oceanic crust is the rock record of a subduction zone. Beneath the folded rocks of the welt, where pressure and temperature are very high, rocks are metamorphosed; foliation is largely vertical or nearly so, mostly at right angles to the horizontal squeezing. As the descending plate reaches depths of extremely high temperatures, part of it melts to generate a silica-rich magma that rises because of its natural buoyancy (it is lighter than the surrounding rock). It intrudes the gneisses and schists and deformed sedimentary rocks to form granitic batholiths (see Chapter 15) or extrudes at the surface as andesitic volcanoes. From this portrayal, we see mountain building is responsible for batholith emplacement, andesitic volcanism, and large-scale metamorphism, folding, and thrust faulting. Examples of continental-oceanic plate mountain belts are the late Mesozoic and early Cenozoic Rockies, the Cenozoic Andes, and the late Paleozoic Appalachians. (When two continental plates converge, little or no subduction occurs because of continental crust buoyancy. The late Paleozoic Urals exemplify this condition. Convergence of two oceanic plates creates a well-developed subduction zone with or without metamorphism and granitic intrusion. This condition is prevalent in the western Pacific with its island arcs.)

When convergence of plates ceases, the mountain belt is temporarily in **isostatic** or flotational **balance** or equilibrium: It floats on denser materials of the lithosphere, and downward depression and thickening of the lower root zone supports the weight of the uplifted mountain range. But uplift of the mountain range causes its immediate erosion, with sediments shed both landward and seaward. And as erosion continues, the isostatic balance is upset. The root zone rebounds or rises to maintain a balance as unloading by erosion occurs at the top of the mountain range. As erosion and isostatic adjustment proceed, the mountain belt becomes thinner, and its upper surface gradually approaches sea level. Erosion has removed the thrust-faulted and folded sedimentary strata, and only metamorphic and intrusive rocks are visible. These are the roots of a once majestic mountain range.

Remember the shields on all continents—central masses of mostly metamorphic and granitic rocks? Well, shields represent, then, roots of former mountain ranges. Shield rocks tend to occur in differently trending structural belts, with the older rocks toward the centers of shields. Inference: Continents grow by the plastering on of mountain belts by plate collision.

Former Continental Seas

Today, a time of high-standing continents, we find it difficult to comprehend seas covering continents. But they did so many times. Present-day, restricted examples are Hudson Bay in North America and the Black and Caspian Seas in southern Europe—all highly unlike some of the extensive continental seas of the past. From sedimentary rocks that reflect the physical, chemical, and biological environments of sediments, geologists interpret seas of the past. Let's use the Late Cambrian rock and fossil record in North America for a glimpse at how the process works.

A sedimentary facies map (Figure 12–3) shows the distribution of major rock bodies. Each **facies,** a distinctive body of rock and associated fossils, reflects a particular sedimentary environment; and each grades laterally and vertically into another. The

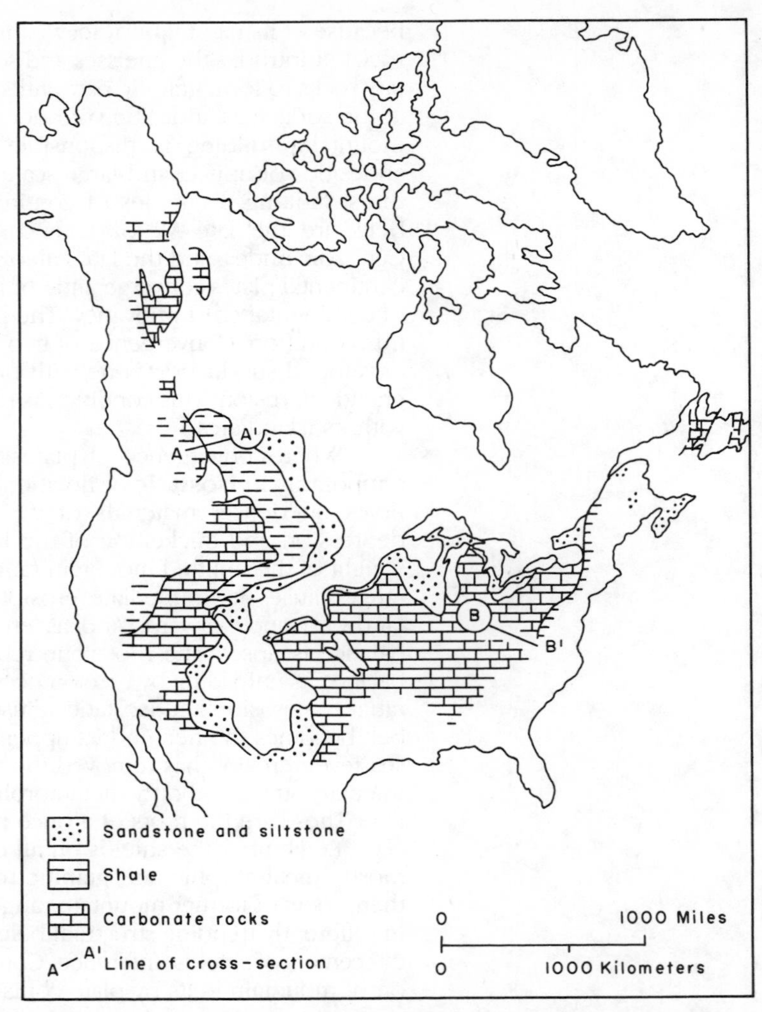

FIGURE 12–3. Sedimentary facies map of Late Cambrian rocks in North America. (Modified considerably from T. D. Cook and A. W. Bally, editors, *Stratigraphic Atlas of North and Central America,* Princeton University Press, 1975, p. 15.) Cross sections relating to this map are in Figure 12–4.

map represents a painstaking compilation of vertical rock sequences from exposures and boreholes made by many geologists over many years, and the correlation of these compilations to produce a unified portrayal. Keep in mind that any spot on the map signifies not exclusively a particular facies but a predominance of that facies.

The sandstone facies, with siltstone, contains common trilobites and brachiopods (see Chapter 18) that attest to its marine origin. Well-sorted and well-rounded sand grains, together with ripple marks and cross-bedding, indicate deposition of the sand-

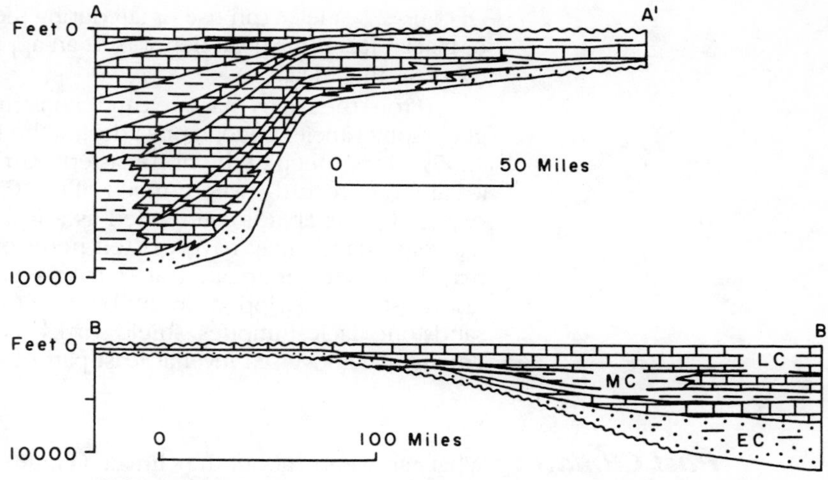

FIGURE 12–4. Cross sections of Cambrian strata—located on Figure 12–3—in western (*A–A'*) and eastern (*B–B'*) North America. Source of the cross sections and rock symbols are the same as for Figure 12–3. *EC*=Early Cambrian, *MC*=Middle Cambrian, *LC*=Late Cambrian.

stone in shallow water and continual reworking by waves and currents. In eastern North America, the sandstone passes directly into a carbonate rock facies containing trilobites, brachiopods, and the algal-related stromatolites. Common clastic carbonates with oölites (see Chapter 16), ripple marks, cross-bedding, and carbonate conglomerate signify considerable shallow-water agitation of limy sediment by waves, currents, and storms. In the west, a shallow-water marine shale intervenes between the sandstone and carbonate rock facies, implying considerable mud deposition there.

Along the margins of the continent, rock sequences thicken (Figure 12–4) because of more downsinking or subsidence as sediments were deposited. The carbonate rock facies gives way outward to a largely deep-water shale facies—containing also rock fragment sandstone and volcanic rocks—with few marine fossils. This facies is poorly known because of later metamorphism and intrusion associated with mountain building.

Cross section *B–B'* (Figure 12–4) shows that Middle Cambrian rocks extend farther toward the center of the continent than Early Cambrian rocks, and Late Cambrian rocks farther still. This establishes a major marine **transgression**—expansion of the sea resulting in submergence of the land—during the Cambrian. It is further reflected in the cross section by offshore rocks—carbonates—overlying nearshore ones—sandstone. What causes a transgression or **regression**—contraction of the sea resulting in emergence of the land—is unclear. Plate tectonics theory suggests that the arching up of an oceanic ridge by abrupt convection could displace sea water from ocean basins onto continents. Diminishing convection allows the oceanic ridge to subside, causing regression.

Of course, sea level can rise or fall during the melting or formation of glacial ice. But glacial deposits of an appropriate age must corroborate this idea.

From the facies map we can reconstruct a map of the paleogeography (ancient geography) during the Late Cambrian (Figure 12–5). Since much of the northeastern part of the continent lacks a Cambrian record—where much of the Canadian Shield is now exposed—this area is interpreted as a low-lying landmass. Alternatively, islands may have existed here; or the entire continent may have been inundated at one time, but the record in the northeast was stripped away by erosion. In any case, the sandstone facies implies shield rocks as its source and their emergence above sea level at least part of the time.

Past Climates

What can we say about the climate (Figure 12–5) during the Late Cambrian in North America? Extensive limestone forms today in low latitudes where higher temperatures result in carbon dioxide release and calcium carbonate precipitation, as in the Bahamas or the Persian Gulf. Reef-building stromatolites may have had similar requirements to reef-building corals, which today occur mostly within 30° of the equator. Our reasoning so far suggests a tropical or subtropical climate during the Late Cambrian.

Now, let's take a different tack. When lava congeals or sediments are deposited, certain iron minerals in the form of crystals or particles—especially magnetite—align themselves parallel to the earth's magnetic field. This happened in the past, with crystals or particles acting as tiny, locked-in compass needles recording past pole positions, providing their magnetism can be measured and interpreted correctly. Inference of pole position by this approach is approximate and is always checked against climatically sensitive rocks. Such paleomagnetic evidence passes the Cambrian paleoequator lengthwise through North America. Drawing in 30° north and south latitudes includes all of the continent within a tropical or subtropical climate, in agreement with reasoning from extensive carbonate rocks and stromatolites.

Applying patterns from present-day climates would place the continent within the easterly trade wind belt, with winds blowing obliquely toward the paleoequator. Such winds would primarily drive the main continental sea currents, probably paralleling sea bottom contours and deflected in places by the irregular shoreline. A generally southerly or southwesterly flow of currents is supported by the orientations of cross-beds (see Chapter 16) in the sandstones.

Although not applicable to our Late Cambrian example, other rock types are useful in generally inferring past climates. Thick, extensive evaporites—such as rock gypsum, rock anhydrite, and rock salt—indicate arid climates causing the high evaporation necessary to allow the salts to precipitate from brines. Dune sand-

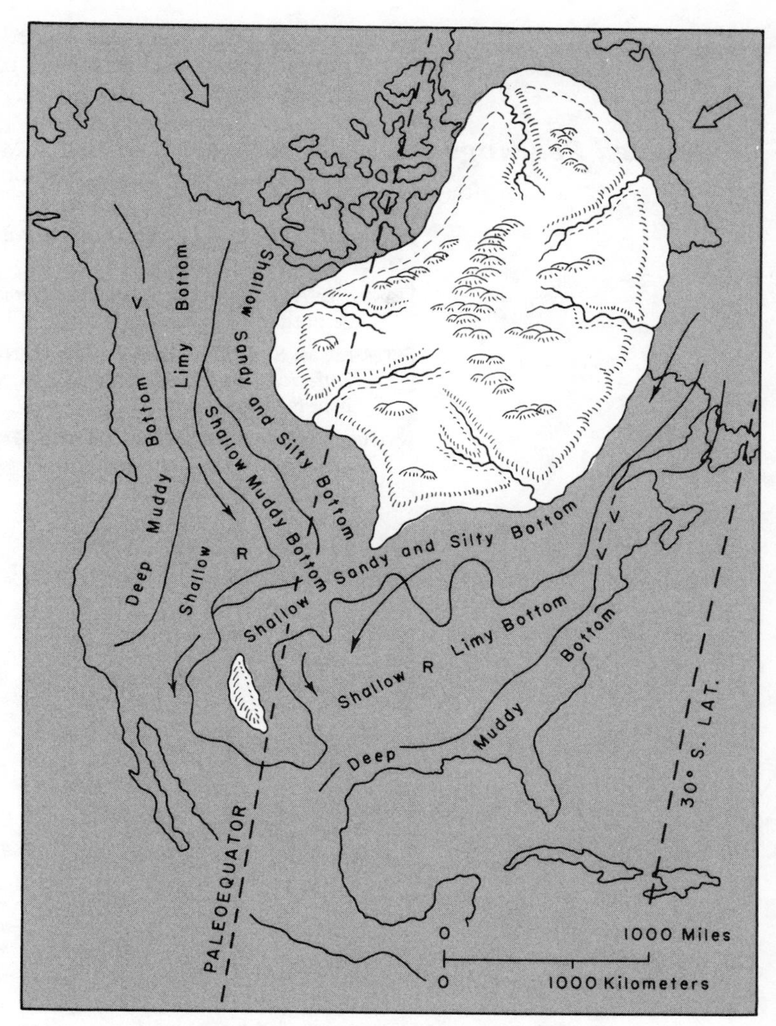

FIGURE 12–5. Reconstruction of the geography and climate during the Late Cambrian. Areas of sea-bottom sediment were derived largely from Figure 12–3. Most of the continent is inferred to have been covered by the sea, except, perhaps, the northeastern part. *R*=stromatolite reefs, *V*=volcanic rocks, narrow arrows=surface oceanic currents, and broad arrows=inferred prevailing winds.

stone, recognized by thick, wedge-shaped intervals of cross-beds, also requires an arid climate for its formation, and the direction of inclined cross-beds gives prevailing wind direction. Thick, extensive coal deposits reflect moist conditions—where rainfall exceeds evaporation—but temperatures may be cool to tropical. Growth rings in preserved wood suggest a temperate, seasonally changing climate; they are lacking if growth occurred within a tropical climate. And, finally, *tillite*—lithified till—is a good indicator of a cool, moist climate compatible with glacier formation. But other poorly sorted rocks may be confused with it. Supposed tillites are best verified if they rest on striated rock surfaces.

Selected Readings

BAMBACH, R. K., C. R. SCOTESE, and A. M. ZIEGLER. "Before Pangea: The Geographies of the Paleozoic World," *American Scientist,* 68 (1980), 26–38.

CONDIE, K. C. *Plate Tectonics and Crustal Evolution* (2nd ed.). New York: Pergamon Press, Inc., 1982.

FRAKES, L. A. *Climates Throughout Geologic Time.* New York: Elsevier Scientific Publishing Co., 1979.

PETERSEN, M. S., J. K. RIGBY, and L. F. HINTZE. *Historical Geology of North America* (2nd ed.). Dubuque, Iowa: William C. Brown Co., Publishers, 1980.

WYLLIE, P. J. *The Way the Earth Works: An Introduction to the New Global Geology and Its Revolutionary Development.* New York: John Wiley and Sons, Inc., 1976.

13 PARKS FOR GEOLOGIC OBSERVATION AND CONTEMPLATION

Why do people visit parks? That's simple. They come to swim, fish, boat, picnic, and just plain relax. But is it that simple? Not really. Many seem compelled occasionally to flee their everyday routine, to seek recreation, to be sure, but particularly to recreate minds as well as bodies. Others may not be so compelled, but simply wish to inform others of having been elsewhere than at their too-familiar domiciles. Many physically fit people come to hike or backpack. Similarly, others arrive for a specific purpose, such as photography. Scientists periodically inhabit parks to research a particular geologic feature or organism. A rare visitor might drop in by accident, perhaps temporarily disoriented or tugged by a gentle inquisitive urge. And others—possibly only a small percentage of visitors—come to learn of nature, followed by a smaller number still of those searching for a natural world, intangible experience or feeling. I may have left out someone. But these are the people and motives readily obvious.

One definition of a park is "any area of public land." In this broad sense, a list of public areas at the national, state, or lesser levels is long and varied. National areas specifically designated include parks, monuments, forests, recreation areas, wilderness areas, seashores, lakeshores, wild and scenic riverways, refuges, scientific reserves, and others. State areas encompass parks, monuments, forests, preserves, natural areas, and others. Many state features or areas derive further sanction as nationally Registered Natural Landmarks. One compilation of numerous natural regions—some privately owned but open to the public—is Hilowitz and Green's *Natural Wonders of America.* County and city parks expand our list even further. Implied so far are natural regions, but many parks are human creations or have been set aside for their historical significance. Now, what about cemeteries—are they parks? Well, they're usually open to the public. Although concentrations of the dead, they can be sources of enjoyment for the living. Cemeteries can be quiet places for relaxation or contemplation. Some people like to watch birds in such places. And others may even observe and investigate the effects of rock weathering (see Chapter 23). Parks, then, exist in all sizes and types at any level, but it's the national parks that tend to really snatch our attention, because of their large size and variety. Let's see how they came about.

On a cool, starry night in mid-September, 1870, four men relaxed before a flickering campfire along the Firehole River in what is now northwestern Wyoming. They were a Surveyor-General for Montana Territory, a second lieutenant in the United States Army, a vigilante law enforcement officer, and a judge for Montana Territory. All were part of an exploring party, with eleven others, that had spent nearly five awesome weeks witnessing such marvels as hot springs, geysers, and glass cliffs. What should be done with such a marvelous place? Perhaps each should stake his claim before word got out. But the judge countered this idea. This area, with its unique beauty and fascinating natural features, must not be fragmented for personal gain. Its uniqueness must be preserved intact for all people of the nation to enjoy—as a national park. They all agreed and set out to promote the idea at the completion of the expedition.

Diligent promotional work paid off. Two years later, eighteen years before Wyoming became a state, Yellowstone arose as the first national park in the United States—and the world. (Yellowstone's first park superintendent was the vigilante law enforcement officer.) It was the largest United States national park—2.2 million acres (0.9 million hectares)—until Wrangell–St. Elias in southern Alaska took the honors as a national monument (now a park) in 1978 with 12.3 million acres (5 million hectares).

How do parks and monuments, the most common federal parks, differ? In a nutshell, parks can be established only by an act of Congress, but monuments need merely a presidential proclamation. And what may be a comforting thought: A park or monument can be abolished only by an act of Congress. An area may be protected sooner if designated a monument and may be elevated later to a park—as was the case for Wrangell–St. Elias. Parks are generally larger and encompass a wider array of natural features. Monuments concentrate on specific natural features or preserve archaeological or historic sites. Mesa Verde, in southwestern Colorado, stands out as an exception: It is a national park established for an archaeological site.

Parks as Geologic Refuges

Most national parks, and many other national areas as well, have been established for their outstanding geologic features: Alaska's Wrangell–St. Elias for its glaciers, Kentucky's Mammoth Cave for its extensive solution cave system, Wyoming's Yellowstone for its unique hydrothermal features, and Arches for its exquisitely sculptured arches (Figure 13–1), to name a few. At the state level, the same is true. A sampling of state parks existing primarily for

FIGURE 13-1. Delicate Arch of sandstone, Arches National Park, southeastern Utah. Person is 5.5 feet (1.7 m) high.

their geological features includes: Connecticut's Dinosaur (dinosaur tracks), Hawaii's Waimea Canyon ("Grand Canyon of the Pacific"), Montana's Lewis and Clark Caverns, South Dakota's Bear Butte (laccolith), and Washington's Ginkgo Petrified Forest. Many other types of natural areas at the state and lower levels preserve geological features.

Many parks, then, are in reality geologic refuges, keeping geological features intact, but offering no protection from the continual forces of natural erosion. They are places where interested persons can come to observe natural beauty enhanced or created by the geology, unadulterated by man. Perhaps, too, they can come to learn geology if that is their desire. Scientists come to study and research geological phenomena, unhampered by possible refused access and altered landscape. These refuges increase in aesthetic and scientific value as developers, extractors of mineral wealth, and others strive to meet the demands of a progressive, competitive society. Of course, geological refuges are biological and archaeological refuges as well, and these factors are often complexly intertwined. We must not attempt to fragment the natural world. But, at times, a specific emphasis is convenient or desirable.

The integrity of a park as a geologic refuge is essentially maintained by rules prohibiting collecting or defacing. Scientists may be allowed to collect mineral, rock, or fossil samples. But this is done only on a restricted basis, generally avoided in highly frequented areas, and must be justified by a projected increase in scientific knowledge.

Using Your Parks Geologically

To those craving a knowledge of the natural world, parks are outdoor museums or—to the scientist—outdoor laboratories. The national park system—including parks, monuments, seashores, and other refuges—can be likened further to a great university, with campuses scattered over the country. Imagine, if you will, selectively visiting a substantial number of these parks (Table 13–1) for their varied geology and devoting a reasonable amount of time to each. Over a period of a few to several years, it's conceivable you may have "earned" the near equivalent of a degree in that field. And you would have gained your knowledge informally, rather effortlessly, and out of the classroom. Granted, you might lack certain higher mathematics, chemistry, and physics requirements, and certain skills in the use of laboratory equipment. But you would have gained more than ample background in biology and perhaps adequate knowledge in arts, social sciences, and humanities, especially if you included a few archaeological and historic parks along the way. (I'm assuming you've polished your communication skills on your own.) You would have grown tall

FIGURE 13–2. Discussing and contemplating geology at day's end by the campfire.

in the appreciation of natural beauty and enriched spiritually in acquiring a reverence for nature's creations.

To whatever extent you wish to learn geology through parks, be prepared to expend some time. You may be familiar with a conversation that goes something like this: "Have you visited Yellowstone?" "Yes, this morning." Everyone has the right to visit a park as he or she wishes, but the longer your stay, the more fulfilling the experience. Visit the park museums—many have them—and acquire some park literature. Attend interpretive programs. Ask questions of park naturalists. If physically able, hike some of the trails, at least the guided nature trails. How your knowledge and appreciation of a park grow by simply leaving the roads, even for a mile or less! And, if at all possible, stay overnight in a park. So much more is gained as you assimilate changing light, sounds, smells, and more geology. At the end of a day further absorb what you've learned, preferably before a campfire (Figure 13–2) where you can discuss, project, question. And contemplate the past.

When should you use your parks geologically? Clearly, when you are so inclined, weather is favorable, and other commitments don't interfere. These days, though, parks are not a very well-kept secret. Some say people are loving their parks to death. Maybe people are overusing parks for one of the other reasons mentioned at the beginning of this chapter. But it's tough to learn geology when rubbing elbows continually with your neighbors. So there might be best times—depending on your viewpoint—if you can accommodate them. My preferences, for temperate latitudes, are early June and September through October. These times tend to relate to ebbs in people—and obnoxious insects.

Table 13-1. UNITED STATES NATIONAL PARKS, NATIONAL SEASHORES, AND NATIONAL MONUMENTS OF GEOLOGICAL INTEREST

PARK (P), MONUMENT (M), OR SEASHORE (S)	*CONSPICUOUS LANDFORMS*[1]	*OF SPECIAL INTEREST*
Alaska		
1. Aniakchak (M)	8	One of the world's largest calderas
2. Denali (P)	3a	Highest mountain—McKinley—in North America
3. Gates of the Arctic (P)	3a	Highest peak in Brooks Range
4. Glacier Bay (P)	3a	Fluctuating tidewater glaciers
5. Katmai (P)	8	Valley of Ten Thousand Smokes
6. Kenai Fjords (P)	3a	Earthquakes frequent
7. Kobuk Valley (P)	2, 5	Dune fields within Arctic Circle
8. Lake Clark (P)	3a	North end of active volcanic Aleutian chain
9. Wrangell–St. Elias (P)	3a	Largest assemblage of glaciers in North America

PARK (P), MONUMENT (M), OR SEASHORE (S)	CONSPICUOUS LANDFORMS[1]	OF SPECIAL INTEREST
Arizona		
10. Canyon de Chelly (M)	2	1000-foot (305 m) sandstone cliffs
11. Chiricahua (M)	2	Sculptured volcanic rocks
12. Grand Canyon (P)	2	Mile (1.6 km)-deep canyon cut in past 9 million years
13. Petrified Forest (P)	2	Silica-filled logs erode from badlands
14. Sunset Crater (M)	8	Crater is cinder cone
Arkansas		
15. Hot Springs (P)	9	Most hot water about 4400 years old
California		
16. Channel Islands (P)	4, 5	Natural oil and gas seeps; marine terraces
17. Death Valley (M)	2	Lowest elevation: −282 feet (−86 m)
18. Devils Postpile (M)	3a	Columns from lava cracked upon cooling
19. Kings Canyon (P)	3a	Exfoliation domes in granite
20. Lassen Volcanic (P)	8	Volcanism in early 1900s
21. Lava Beds (M)	8	Lava tubes and caves
22. Pinnacles (M)	2	Sculptured volcanic rocks
23. Point Reyes (S)	4	San Andreas fault zone separates seashore from mainland
24. Redwood (P)	4, 7	Seismic sea wave damage at north end
25. Sequoia (P)	3a	Exfoliation domes in granite
26. Yosemite (P)	3a	Exfoliation domes in granitic rocks
Colorado		
27. Black Canyon of the Gunnison (M)	2	Half mile (0.8 km)-deep canyon cut in past 2 million years
28. Colorado (M)	2	Mesozoic rocks directly overlie Precambrian
29. Dinosaur (M)	2, 9	Dinosaur quarry in Jurassic sandstone
30. Florissant Fossil Beds (M)	2	Fossil fish, leaves, and insects in lake beds
31. Great Sand Dunes (M)	5	700-foot (213 m)-high transverse and barchan dunes
32. Mesa Verde (P)	2	Mesa Verde is cuesta
33. Rocky Mountain (P)	3a	Longs Peak higher than 14,000 feet (4267 m)

PARK (P), MONUMENT (M), OR SEASHORE (S)	CONSPICUOUS LANDFORMS[1]	OF SPECIAL INTEREST
Florida		
34. Biscayne (P)	4	Most northerly coral reefs in U. S.
35. Canaveral (S)	4, 5	White beaches
36. Everglades (P)	2, 4	Coastal swamps protect shorelines against hurricanes
Hawaii		
37. Haleakala (P)	8	Latest eruption about 1790
38. Hawaii Volcanoes (P)	8	Mauna Loa volcano rises 6 miles (9.7 km) above sea floor
Idaho		
39. Craters of the Moon (M)	8	Lava from 40-mile-long (64 km) Great Rift
Kentucky		
40. Mammoth Cave (P)	6	Longest cave system in world
Maine		
41. Acadia (P)	3b, 4	Only true fiord on U. S. Atlantic Coast
Maryland		
42. Assateague Island (S)	4, 5	Assateague is barrier island
Massachusetts		
43. Cape Cod (S)	4, 5	Curved spit at north end of Cape
Michigan		
44. Isle Royale (P)	9, 3b	Topographic "grain" parallels length of island
45. Voyageurs (P)	3b	Rocks are roots of Precambrian mountains
Mississippi		
46. Gulf Islands (S)	4, 5	Barrier islands Mississippi to Florida
Montana		
47. Glacier (P)	3a	Precambrian rocks thrust over Cretaceous
Nebraska		
48. Agate Fossil Beds (M)	2	Cenozoic mammals
Nevada		
49. Lehman Caves (M)	6	Flank of 13,000-foot (3,982 m) Wheeler Peak
New Mexico		
50. Capulin (M)	8	Capulin is cinder cone
51. Carlsbad Caverns (P)	6	Caverns mostly in limestone of Permian barrier reef

PARK (P), MONUMENT (M), OR SEASHORE (S)	CONSPICUOUS LANDFORMS[1]	OF SPECIAL INTEREST
New Mexico		
52. White Sands (M)	5	Dunes of gypsum
New York		
53. Fire Island (S)	4, 5	Barrier island
North Carolina		
54. Cape Hatteras (S)	4, 5	Seashore is barrier island
55. Cape Lookout (S)	4, 5	Three barrier islands
56. Great Smoky Mountains (P)	2	In highest part of eastern U. S.
North Dakota		
57. Theodore Roosevelt (P)	2	Badlands cut into lignite-bearing Paleocene rocks
Oregon		
58. Crater Lake (P)	8	Deepest lake in U. S. occupies caldera
59. John Day Fossil Beds (M)	2	Cretaceous and Cenozoic fossils
60. Oregon Caves (M)	6	Caves in marble
South Dakota		
61. Badlands (P)	2	Badlands cut into Oligocene and Miocene rocks
62. Jewel Cave (M)	6	Named after jewel-like calcite crystals
63. Wind Cave (P)	6	Air currents blow through cave
Tennessee		
Great Smoky Mountains (see North Carolina)		
Texas		
64. Big Bend (P)	2, 9	South-bounded by Big Bend of Rio Grande River
65. Guadalupe Mountains (P)	2	Most of mountains capped by Permian barrier reef
66. Padre Island (S)	4, 5	Barrier island backed by Laguna Madre
Utah		
67. Arches (P)	23, 2	Most natural arches in U. S.
68. Bryce Canyon (P)	23, 2	Joints control location of pillars and arches
69. Canyonlands (P)	23, 2	Joints control location of fins, arches, pillars

PARK (P), MONUMENT (M), OR SEASHORE (S)	CONSPICUOUS LANDFORMS[1]	OF SPECIAL INTEREST
Utah		
70. Capitol Reef (P)	2, 23	"Capitol" for erosional domes, "Reef" for escarpment
71. Cedar Breaks (M)	2, 23	"Cedar" for juniper, "Breaks" for cliffs
72. Natural Bridges (M)	2, 23	Three natural bridges in marine sandstone
73. Rainbow Bridge (M)	2	World's largest natural bridge
74. Timpanogos Cave (M)	6	Three caverns along fault zones
75. Zion (P)	2	1500-foot (457 m) sandstone cliffs
Virginia		
Assateague Island (see Maryland)		
76. Shenandoah (P)	2	Astride crest of Blue Ridge Mountains
Virgin Islands		
77. Virgin Islands (P)	2, 4	Luxurious living coral reefs
Washington		
78. Mount Ranier (P)	3a	Dormant composite volcano in Cascades
79. North Cascades (P)	3a	More than 300 active glaciers
80. Olympic (P)	3a, 4	60 active glaciers; hot springs along faults
Wyoming		
81. Devils Tower (M)	2	Intrusive with columnar structure
82. Fossil Butte (M)	2	Cenozoic fish
83. Grand Teton (P)	3a	Teton Range is upfaulted block
84. Yellowstone (P)	8	Numerous hydrothermal features

[1]Landform numbers are those of chapters as follows:

2 = stream-related
3a = valley glacier-related
3b = continental glacier-related
4 = shoreline-related
5 = wind-related
6 = ground water-related
7 = landslide-related
8 = volcanic
9 = rock deformation-related
23 = weathering-related

Selected Readings

HARRIS, ANN, and ESTHER TUTTLE. *Geology of the National Parks* (3rd ed.). Dubuque, Iowa: Kendall/Hunt Publishing Company, 1983.

HARRIS, D. V. *The Geologic Story of the National Parks and Monuments* (3rd ed.). New York: John Wiley and Sons, 1980.

HILOWITZ, BEVERLEY, and S. E. GREEN. *Natural Wonders of America.* New York: American Publishing Co., 1980.

National Geographic Society Editors, *The New America's Wonderlands, Our National Parks* (rev. ed.). Washington, D. C.: National Geographic Society, 1980.

Minerals, rocks, and fossils: the stuff of geology

14 MINERALS

What Are Minerals?

Minerals are basic components of the earth's crust. More specifically, they are *naturally occurring, inorganic substances with a definite chemical composition and characteristic physical properties.* This definition is quite a mouthful. Let's dissect it and see what it really means.

Naturally occurring means, of course, that a substance is a product of nature, not man-made. Ruby and quartz, for example, have been synthesized in the laboratory; such products may closely resemble their true mineral counterparts, but they are *synthetic,* not *natural* minerals.

Inorganic substance signifies one lacking an organic make-up and formed by inorganic processes. (*Organic* refers to living or once-living things—mainly made of carbon, hydrogen, and oxygen.) Most consider such a substance a solid—not a liquid or gas—but unanimous agreement on this point is lacking.

Definite chemical composition is a stricter description than is commonly true. Minerals are, in fact, chemicals—either elements or compounds. **Elements** consist of a single type of **atom**—a minute particle that is the basic building block of all matter. **Compounds** are made up of more than one kind of atom; most minerals are compounds. Now, the composition of a few minerals is reasonably "definite": Quartz, for example, consists of silicon (Si) and oxygen (O), expressed in chemical shorthand as SiO_2. But for most minerals, atom types other than the primary ones may be present, and one atom type may substitute for another. So the "definite" chemical composition usually *ranges* within certain limits; and, generally, the simpler the composition, the less its variation.

Physical properties are those nonchemical characters—generally usable in the field—that typify all minerals. It is these characters that can generally be used solely to identify the most common minerals, and they will be described shortly. The study of minerals is **mineralogy.**

As you can see, geologists use the word *mineral* in a different, more specific sense than the general population. Others may consider all substances as either "animal, plant, or *mineral*," or say a food is rich in "vitamins and *minerals*."

Minerals are usually grouped by their chemical composition. Table 14-1 chemically groups those minerals in the mineral identification tables (Tables 14-4 to 14-6). Graphite, like gold, for example, occurs in the uncombined state, and is considered a native element. Pyrite, on the other hand, a sulfide, consists of the metallic element iron combined with sulfur; and orthoclase, a silicate, consists of potassium and aluminum combined with silicon and oxygen.

About 2500 minerals are known, and more are discovered

Table 14-1. COMMON MINERAL GROUPS WITH MINERAL EXAMPLES FROM THOSE IN TABLE 14-2

MINERAL GROUP	*GENERAL COMPOSITION*	*MINERAL EXAMPLES*
Native elements	Single element	Graphite
Sulfides	Element(s) + sulfur	Chalcopyrite Galena Pyrite Sphalerite
Oxides	Elements(s) + oxygen	Corundum Hematite Limonite Magnetite
Halides	Element(s) + chlorine or fluorine	Fluorite Halite
Carbonates	Element(s) + carbon and oxygen	Calcite Dolomite
Sulfates	Element(s) + sulfur and oxygen	Anhydrite Gypsum
Phosphates	Element(s) + phosphorus and oxygen	Apatite
Silicates	Element(s) + silicon and oxygen	Amphibole Biotite Chlorite Garnet Muscovite Olivine Orthoclase Plagioclase Pyroxene Quartz Serpentine Staurolite Talc

Table 14-2. **COMPOSITION AND SELECTED USE OF COMMON MINERALS**

MINERAL	*COMPOSITION*	*SELECTED USE(S)*
Amphibole	Hydrous calcium, magnesium, iron, aluminum silicate	Rock former
Anhydrite	Calcium sulfate	Soil conditioner
Apatite	Calcium, fluorine phosphate	Source of fertilizer
Biotite	Potassium, magnesium, iron, aluminum silicate	Rock former
Calcite	Calcium carbonate	Making of cement and lime (from limestone)
Chalcopyrite	Copper iron sulfide	Important ore of copper
Chlorite	Hydrous magnesium, iron, aluminum silicate	Rock former
Corundum	Aluminum oxide	Abrasive and gemstone
Dolomite	Calcium magnesium carbonate	Making of cement and building stone (from dolostone)
Fluorite	Calcium fluoride (halide)	Flux in steel-making; making of hydrofluoric acid
Galena	Lead sulfide	Chief ore of lead
Garnet	Calcium, iron, magnesium, aluminum silicate	Gemstone and abrasive
Graphite	Carbon	Lubricant; making of "lead" pencils
Gypsum	Hydrous calcium sulfate	Making of plaster of Paris
Halite	Sodium chloride (halide)	Source of salt, sodium, and chlorine
Hematite	Iron oxide	Chief ore of iron
Limonite	Hydrous iron oxide	Ore of iron
Magnetite	Iron oxide	Important ore of iron
Muscovite	Potassium aluminum silicate	Electrical insulation
Olivine	Magnesium iron silicate	Minor gemstone; refractory material for casting
Orthoclase	Potassium aluminum silicate	Making of ceramics
Plagioclase	Sodium calcium aluminum silicate	Making of ceramics
Pyrite	Iron sulfide	Source of sulfur for sulfuric acid
Pyroxene	Calcium, magnesium, iron, aluminum silicate	Rock former
Quartz	Silicon dioxide	Making of glass; gemstone
Serpentine	Hydrous magnesium silicate	Chief source of asbestos (chrysotile)
Sphalerite	Zinc sulfide	Chief ore of zinc
Staurolite	Iron aluminum silicate	Minor gemstone
Talc	Hydrous magnesium silicate	Making of talcum powder

continually. Of this number, though, only about 20 to 30 are common, and about 10 or so—all silicates—constitute more than 90 percent of the earth's crust.

Minerals are of major importance to humans, and have been since prehistoric time. Several contain valuable metals—such as iron, copper, and lead—and are classed as **ore minerals.** (An **ore** is a mineral or group of minerals that can be mined at a profit.) Common examples are chalcopyrite, galena, and hematite (Table 14–2). Others, **industrial minerals,** are utilized in the raw state or appropriately processed, but a metal is not extracted from them. Examples of this group are fluorite, graphite, and talc.

Now, what is the relationship between minerals and rocks? Simply, minerals make up rocks, or, **rocks** *are aggregates of minerals.* This simple definition, however, doesn't always hold true. In certain cases, a rock may consist of only one mineral, such as calcite making up the rock limestone. Earlier, we defined minerals as *inorganic* substances, so the rocks of which they are composed should be inorganic as well. But coal consists largely of organic, decomposed plant matter, although inorganic minerals may be admixed as well. Irrespective of its predominantly organic make-up, coal qualifies as a rock because it occurs naturally and often contains included minerals.

While we're on the subject of exceptions, what about water (or ice) and oil? Water has a rather definite chemical composition—two parts hydrogen plus one part oxygen—although impurities are frequent. So it can be considered a mineral. Purists, however, would say only a solid—such as water in the form of snow or ice—can be a mineral. This approach would exclude liquid mercury as well. Oil has the dual problem of having an organic composition, as does coal, and liquid form. Its highly variable composition might support the contention that this hydrocarbon is a rock. Usual practice, however, is to term oil a *mineral fuel.*

Identifying Minerals

Sophisticated laboratory methods are required for positive identification of most minerals. Included are the use of X-rays, chemical analysis, and analysis of optical properties. The last approach involves cutting and grinding a mineral to paper thinness and observing through a microscope how it behaves as ordinary and polarized light pass through it. For the most common minerals, however, the *physical properties* generally serve to identify them. Properties used primarily in this chapter are luster, color and streak, hardness, cleavage and fracture, crystal form, and heft or specific gravity.

Luster

The appearance of a mineral in reflected light is its **luster.** It is of two types: metallic and nonmetallic. Metallic luster is that resembling a shiny metal (e.g., Figure 14–3); minerals with this luster are opaque and do not transmit light. Nonmetallic luster—present in minerals that transmit light, at least through thin edges—is varied; it includes glassy, resinous (as in resin), pearly, greasy (as of oily glass), silky (as of a fibrous mineral), brilliant (as in diamond), and dull (no luster).

Color and streak

Color is the most obvious physical property of a mineral. In some minerals, as in the metallic ones, it is constant and useful for identification. In others, as in quartz, color varies widely. Always discern color on fresh surfaces, since minerals are frequently stained or tarnished. "Light-colored" (Table 14–5) generally means colorless, white, yellow, orange, light red, light brown, light gray, light green, and light blue. Other colors can be considered "dark" (Table 14–6). **Streak,** the color of the powdered mineral, is more diagnostic than a mineral's color in reflected light. For example, hematite may be reddish brown to black but always produces a brownish red streak. Streak is most useful for the dark, especially metallic, minerals. Streak can be obtained in three ways: rubbing a mineral against unglazed porcelain (easiest), rubbing two pieces of the same mineral together, or attempting to cut a groove in a mineral with a knife blade. You may lack access to unglazed porcelain, but with a little forethought you could acquire a piece of ceramic bathroom tile (back side), which serves equally well. For the last two methods, try to collect the powdered mineral on paper or a light surface to readily ascertain its color.

Hardness

A mineral's resistance to scratching is its **hardness.** Certain common minerals are used in a scale of hardness (Table 14–3). Minerals with large numbers in the scale can scratch those with smaller index numbers. Diamond is the hardest natural substance known. The hardness of a mineral is controlled by its internal atomic structure. Harder minerals have a stronger bonding of atoms or ions—electrically charged atoms—smaller atoms (which thus fit more tightly), or more closely packed atoms.

If a mineral hardness set is not readily accessible, other common materials—a fingernail, copper penny, knife blade, window glass, and steel file (Table 14–3)—will work. Before attempting a hardness test, make sure the mineral surface selected is fresh and unweathered. Start with a knife blade. (It's a good idea to carry a small folding pocketknife with two or three blades. Reserve one blade for hardness tests only so all don't dull frequently.) Hold the mineral firmly and attempt to scratch a groove in the mineral with the knife blade. Use considerable pressure for a meaningful test. Blow away any ground mineral or powder and feel for a groove with your fingernail or examine the mineral with a hand lens. If a definite groove exists, the mineral is softer than your knife blade. You may then refine the hardness value by testing the mineral with a copper penny or your fingernail. If the mineral is harder than the knife blade, a faint metallic streak from the knife blade may be left on the mineral.

Table 14-3. STANDARD SCALE OF MINERAL HARDNESS AND HARDNESS OF COMMON MATERIALS

HARDNESS INDEX	*MINERALS*	*COMMON MATERIALS*
1	Talc (softest)	
2	Gypsum	
2.5		Fingernail
3	Calcite	Copper penny
4	Fluorite	
5	Apatite	
5.5		Knife blade, window glass
6	Orthoclase	
6.5		Steel file
7	Quartz	
8	Topaz	
9	Corundum	
10	Diamond (hardest)	

Cleavage and fracture

Cleavage is the ability of a mineral to break along smooth, flat planes. A mineral may break along one to six directions. Biotite, for example, has cleavage in one direction and breaks into thin sheets (see Figure 14–19). Galena, on the other hand, has three directions of cleavage at right angles to each other and so breaks readily into cubes (Figure 14–2). Cleavage, like hardness, is controlled by internal atomic structure. It occurs along a direction of weakness; that is, a direction across which the atomic bonding is relatively weak. When a mineral is rotated in reflected light, a "flash" of light suggests a cleavage surface. Small steps on a mineral's surface or parallel cracks—in transparent minerals—also indicate cleavage (Figures 14–7, 14–9).

Fracture is the ability of a mineral to break along surfaces other than smooth, flat planes; preferred directions of weakness, as in cleavage, are not evident. Common types of fracture are *shell-like* or *conchoidal,* a smooth, curved fracture resembling the inside of a clamshell; *uneven* or *irregular;* and *splintery* or *fibrous.*

Crystal form

A **crystal** (e.g., Figure 14–30) is a symmetrical, geometric form bound by smooth, flat surfaces—crystal faces—reflecting ordered internal atomic structure. Nearly all minerals are crystalline—have an ordered internal arrangement of atoms—but they commonly lack an external crystal form. Crystals form from solutions, melts (melted substances), and vapors. You can acquire a good idea of how crystals form and grow if you have access to a microscope. Place a drop of a concentrated solution of table salt and water on a glass slide. As the water evaporates, ions of sodium (Na^+) and chlorine (Cl^-) combine to form eight-faced crystals—first along the margin of the drop. Snowflakes—six-sided crystals—are well-known examples of crystallization from a (water) vapor.

The study of crystals is rather complex. Here, we will consider only a few common crystal forms (Tables 14–4 to 14–6) to help identify the common minerals.

Be careful not to confuse a crystal face with a cleavage plane. A crystal face has only one other face parallel to it and may portray growth lines (Figure 14–4). A cleavage plane has an unlimited number of potential parallel planes. Crystal faces may (Figure 14–13) or may not (Figures 14–8, 14–10) parallel cleavage planes.

Heft or specific gravity

Heft simply means how heavy a mineral feels when lifted in the hand. Technically, it is *specific gravity,* or the ratio of the weight of a mineral to the weight of an equal volume of water. Quartz, with a specific gravity of 2.6, weighs 2.6 times that of water. Minerals weighing less than quartz can be considered "light," those weighing more are "heavy." Galena and pyrite are heavy minerals. Light and heavy minerals can be readily sensed with a little "hefting" practice.

Other properties

Observation of other physical properties may be occasionally useful in mineral identification. These include feel (e.g., greasy, as in graphite), attraction to a magnet (e.g, present in magnetite), taste (e.g., salty in halite), bubbling when hydrochloric acid is added (e.g., calcite), and being flexible in thin layers (e.g., chlorite) or flexible and elastic (e.g., biotite). If you intend to carry an eyedropper bottle with dilute acid, mix one part hydrochloric acid with five parts water. CAUTION: Always add acid to water, *never* water to acid.

Process of Identification

Identification of the common minerals is accomplished by evaluating the physical properties just examined in the case of an unknown mineral and using the mineral identification tables (Tables 14–4 to 14–6) and illustrations. Let's take a mineral in hand and see how the identification process works.

Our pale blue mineral has glassy luster, so we go to Table 14–5. The knife scratches it rather easily, leaving behind a white powder. Good cleavage is present, evidenced by several steps on the mineral's surface, presumably in three directions at oblique angles. This results in rhombohedron cleavage forms, shaped somewhat like a weakened, skewed wooden box after someone sat on one corner. Checking the hardness further, we find it to be about that of a penny. The combination of characters steers us to calcite or dolomite. If we have dilute hydrochloric acid with us, we can finalize the identification to calcite by noting rapid bubbling. If no acid is available, we might try lemon juice. Yes, lemon juice! This contains very weak acid, so any bubbling must be viewed carefully with a hand lens. (Vinegar contains a weak acid too, and may be used similarly to lemon juice.) You might wonder: Isn't this a bit deceiving—*blue* calcite? Perhaps so. But I wished to make a point. Don't place too much emphasis on color. Rely more on other properties.

There you have it; that's the basic identification process. If you are unable to identify a mineral with one table, try another in case you made a wrong decision somewhere. If you still cannot identify the mineral, try one of the identification manuals listed at the end of this chapter. Remember, though, only a relatively few minerals can be identified by their physical properties. You may, if you become really serious about minerals, have to resort to chemical tests or consult a specialist at a university or museum.

Selected Readings

CHESTERMAN, C. W. *The Audubon Society Field Guide to North American Rocks and Minerals.* New York: Alfred A. Knopf, Inc., 1978.

MOTTANA, A., R. CRESPI, and G. LIBORIO. *Guide to Rocks and Minerals.* New York: Simon and Schuster, Inc., 1978.

SORRELL, C. A. *Rocks and Minerals.* New York, Golden Press, 1973.

ZIM, H. S., and P. R. SHAFFER. *Rocks and Minerals, a Guide to Familiar Minerals, Gems, Ores and Rocks.* New York: Golden Press, 1957.

Table 14-4. MINERAL IDENTIFICATION: METALLIC LUSTER

SOFTER THAN KNIFE	
Cleavage Good	
GRAPHITE.	Color and streak black; softer than fingernail, smudges fingers; cleavage in one direction; six-sided, tabular crystals, but usually in foliated or scaly masses; greasy feel (Figure 14–1).
GALENA.	Color and streak lead-gray; about as hard as fingernail; cubic cleavage and crystals; heavy (Figure 14–2).
Cleavage Poor or Absent	
CHALCOPYRITE.	Brass-yellow; often tarnished to bronze or iridescent; streak greenish black; harder than penny; fracture uneven; usually in masses; brittle; also called "fool's gold," as is pyrite (Figure 14–3).
HARDER THAN KNIFE	
Cleavage Poor or Absent	
PYRITE.	Pale brass-yellow, darker if tarnished; streak greenish or brownish black; about as hard as steel file; fracture shell-like; cubic crystals most common, also in masses and grains; heavy; brittle; also called "fool's gold," as is chalcopyrite (Figure 14–4).
HEMATITE (crystalline variety).	Dark brown to black; streak brownished red; fracture shell-like; crystals equidimensional to tabular, also in rounded and scaly masses; heavy (Figure 14–5).
MAGNETITE.	Color and streak black; about as hard as steel file; fracture uneven; crystals often eight-faced, also commonly in grains or masses; heavy; attracted to magnet, may also act as magnet (*lodestone*) (Figure 14–6).

FIGURE 14–1. Graphite, from Columbo, Ceylon, showing metallic luster and easily scratchable surfaces. Width of specimen is 4.2 inches (10.7 cm).

FIGURE 14–2. Galena, showing metallic luster and cubic cleavage. Smallest cleavage fragment is reflected in cleavage surface of intermediate-sized fragment. Height of largest specimen is 2.2 inches (5.6 cm).

FIGURE 14–3. Chalcopyrite, showing high metallic luster. Height of specimen is 2.0 inches (5.1 cm).

FIGURE 14–4. Cubic crystal of pyrite, showing metallic luster and shell-like fracture and vertical growth lines on front crystal face. Height of specimen is 2.2 inches (5.6 cm).

FIGURE 14–5. Hard, massive variety of hematite, from Minnesota, showing metallic luster and rounded surfaces. Width of specimen is 3.5 inches (8.9 cm).

FIGURE 14–6. Massive magnetite, from Baltimore, Maryland. Bar magnet attracted to it is 0.9 inch (2.3 cm) high.

Table 14-5. **MINERAL IDENTIFICATION: NONMETALLIC LUSTER, LIGHT-COLORED**

SOFTER THAN KNIFE

Cleavage Good	
HALITE.	Luster glassy; colorless or white, other colors if impure; about as hard as fingernail; cleavage cubic; crystals cubic, also in masses and grains; light; salty taste (Figure 14-7).
FLUORITE.	Luster glassy; color variable, most often yellow, green, or purple; harder than penny; cleavage in four directions, parallel to faces of octahedron; crystals usually cubic, often intergrown, also in masses (Figure 14-8).
CALCITE.	Luster glassy to dull; colorless or white, other colors if impure; hard as penny; cleavage in three directions, at oblique angles; crystal form highly varied, rhombohedrons (each face a rhombus) and scalenohedrons (each face a scalene triangle, "dog-tooth" type) most common, also in masses and grains; bubbles rapidly in dilute hydrochloric acid (Figures 14-9, 14-10).
DOLOMITE.	Luster glassy to pearly; color variable, often pink or flesh-colored; harder than penny; cleavage as in calcite; crystals commonly curved rhombohedrons (see for calcite), also in masses and grains; bubbles slowly in dilute hydrochloric acid, rapidly when mineral is powdered; less common than calcite (Figure 14-11).
ANHYDRITE.	Luster glassy to pearly; colorless, white, gray, or variously tinted; about as hard as penny; cleavage in three directions, yields cubic or rectangular fragments; crystals rare, usually in masses; often alters to gypsum by addition of water (Figure 14-12).
GYPSUM.	Luster glassy, pearly, or silky; colorless white, gray, or variously tinted; softer than fingernail; good cleavage in one direction; crystals mostly tabular, commonly intergrown ("fishtail"), also in masses or fibrous; varieties include *selenite* (colorless, crystalline), *satin spar* (fibrous), and *alabaster* (massive) (Figure 14-13).
TALC.	Luster pearly to greasy; commonly green, white, or gray; much softer than fingernail; cleavage in one direction, thin layers flexible but not elastic; crystals rare, usually sheetlike or in compact masses (soapstone); greasy or soapy feel (Figure 14-14).
MUSCOVITE.	Luster glassy to pearly; colorless (thin sheets) to yellow brown, green, and red; about as hard as fingernail; cleavage in one direction, thin layers flexible and elastic; crystals tabular, often six-sided, also in sheetlike and scaly masses; one of the micas (Figure 14-15).

HARDER THAN KNIFE

Cleavage Good	
ORTHOCLASE.	Luster glassy to pearly; usually white, gray or flesh-colored; about as hard as steel file; cleavage in two directions, at right angles; crystals usually short prisms, often intergrown, also in cleavable or granular masses; one of the feldspars (Figure 14-16).
PLAGIOCLASE.	Similar to orthoclase but usually white or gray and has fine parallel lines on good cleavage surfaces; one of the feldspars.

Cleavage Poor or Absent

QUARTZ.	Luster glassy; usually colorless or white, but may be any color; about as hard as steel file or harder; fracture shell-like; crystals usually as prisms that appear to be capped by pyramids, also in masses; many varieties, including *rock crystal* (colorless) (Figure 14–17), *milky quartz* (white), *rose quartz* (pink) (Figure 14–18), *amethyst* (purple), *smoky quartz* (smoky yellow to black), and *agate* (banded or mossy).

FIGURE 14–7. Halite, from Windsor, Ontario, showing glassy luster and cubic cleavage. Height of specimen is 3.6 inches (9.1 cm).

FIGURE 14–8. Intergrown cubic crystals of fluorite, showing glassy luster. Straight cracks in some crystals reflect cleavage surfaces that are oblique to crystal faces. Width of specimen is 2.6 inches (6.6 cm).

FIGURE 14–9. Cleavage fragment of calcite, from Riverside, California, showing glassy luster and three directions of cleavage. Width of left front part of specimen is 3.7 inches (9.4 cm).

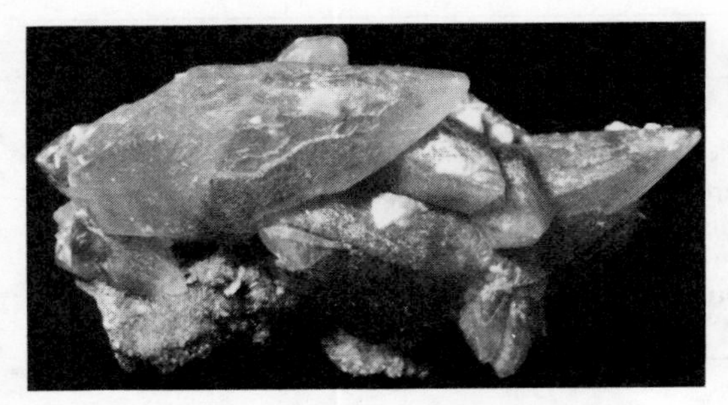

FIGURE 14-10. "Dog-tooth" crystals of calcite, from Ohio. Uppermost crystal displays a crack that reflects a cleavage surface oblique to crystal faces. Width of crystal group is 4.6 inches (11.7 cm).

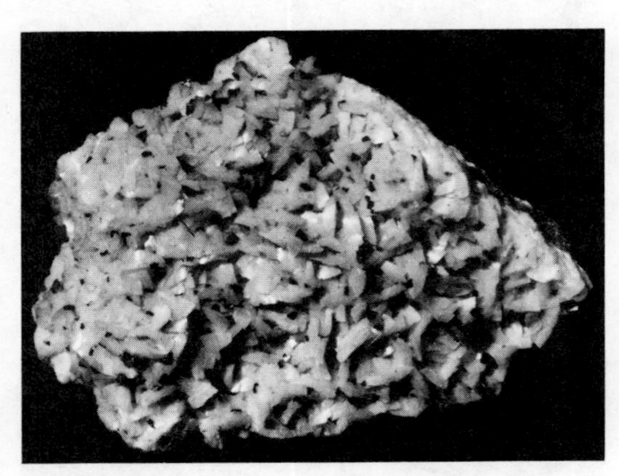

FIGURE 14-11. Curved crystals of dolomite—with smaller, darker crystals of chalcopyrite—from Joplin, Missouri, showing pearly luster. Height of specimen is 2.6 inches (6.6 cm).

FIGURE 14-12. Massive anhydrite, showing glassy to pearly luster. Width of specimen is 3.1 inches (7.9 cm).

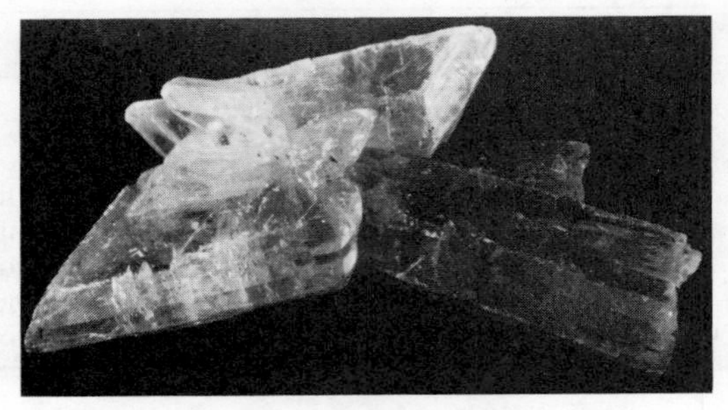

FIGURE 14-13. Intergrown crystals of gypsum, from southwestern North Dakota, showing glassy luster. Straight crack in lower left of specimen reflects a cleavage surface that parallels a crystal face. Width of specimen is 4.3 inches (10.9 cm).

FIGURE 14-14. Cleavable mass of talc, showing pearly to greasy luster. Width of specimen is 5.1 inches (12.9 cm).

FIGURE 14-15. Muscovite mica, from Keystone, South Dakota, showing the capability of cleaving into thin sheets. Height of specimen is 6.1 inches (15.5 cm).

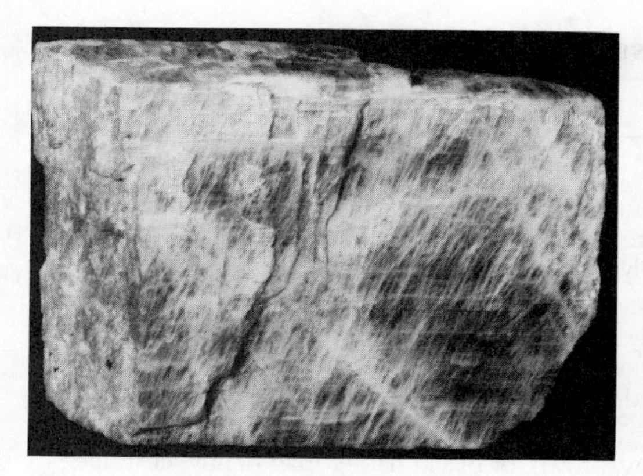

FIGURE 14-16. Orthoclase feldspar, from Delaware County, Pennsylvania, showing two directions of cleavage at right angles, revealed by front and top surfaces, and pearly luster. Height of specimen is 3.1 inches (7.9 cm).

FIGURE 14-18. Massive, noncrystalline, rose quartz, from Hot Springs, Arkansas. Height of specimen is 4.2 inches (10.7 cm).

FIGURE 14-17. Transparent, six-sided crystal prisms of quartz capped by pyramids, showing glassy luster. Height of largest crystal is 1.8 inches (4.6 cm).

Table 14-6. MINERAL IDENTIFICATION: NONMETALLIC LUSTER, DARK-COLORED

SOFTER THAN KNIFE	
Cleavage Good	
BIOTITE.	Luster glassy to pearly; usually dark green, brown, or black; hard as fingernail or penny; cleavage in one direction, thin layers flexible and elastic; crystals tabular, often six-sided, usually in sheetlike or scaly masses; one of the micas (Figure 14–19).
CHLORITE.	Luster glassy to pearly; usually green to blackish green; hard as fingernail or somewhat softer; cleavage in one direction, thin layers flexible but not elastic; crystals tabular, often six-sided, usually in sheetlike or scaly masses (Figure 14–20).
SPHALERITE.	Luster resinous; usually yellow, brown, or black; harder than penny; cleavage in six directions; crystals in tetrahedrons common, usually in cleavable masses (Figure 14–21).
Cleavage Poor or Absent	
SERPENTINE.	Luster waxy, greasy, or silky; usually green, but variable; usually harder than penny; fracture shell-like or splintery; no crystals, usually in masses (Figure 14–22) or fibers (Figure 14–23) (fibrous variety is *chrysotile* asbestos); smooth or greasy feel.
APATITE.	Luster glassy to resinous; usually green or brown; harder than penny; fracture shell-like; crystals of six-sided prisms, also in masses or grains (Figure 14–24).
LIMONITE.	Luster dull to glassy; yellow, yellow-brown to black; streak yellow-brown; softer than fingernail to about that of knife; fracture shell-like to uneven; no crystals, in earthy or compact, rounded masses (Figure 14–25).
HEMATITE (earthy variety).	Luster dull; red and reddish brown; streak brownish red; hardness softer than fingernail to about that of knife; fracture uneven; no crystals, in earthy masses; heavy.
HARDER THAN KNIFE	
Cleavage Good	
PYROXENE.	Luster glassy to pearly; usually gray, green or black; cleavage in two directions, at about right angles, forming square or rectangular cleavage fragments; crystals commonly stubby, eight-sided prisms, also in compact masses and grains; a group of minerals, most common is *augite* (Figure 14–26).
AMPHIBOLE.	Luster glassy to silky; usually gray, green, or black; cleavage in two directions, not at right angles, forming wedge-shaped cleavage fragments; crystals commonly long, six-sided prisms, also in fibrous or irregular masses; a group of minerals, most common is *hornblende* (Figure 14–27).
STAUROLITE.	Luster glassy to dull; brown to black; harder than steel file; fair cleavage in one direction; usually in crystals—stubby prisms—often intergrown as crosses (staurolite from Green *stauros*, "cross") (see Figure 17–4).
Cleavage Poor or Absent	
CORUNDUM.	Luster glassy to diamondlike; color variable, usually brown, gray, red, or blue; harder than steel file; fracture uneven to shell-like; crystals six-sided, commonly barrel-shaped or tabular, also in grains; gem varieties include *ruby* (red) and *sapphire* (blue) (Figure 14–28).

Cleavage Poor or Absent

OLIVINE.	Luster glassy; "olivine is olive green" (usually); hard as steel file or harder; fracture shell-like; usually in grainy masses (Figure 14–29).
GARNET.	Luster glassy to resinous; usually red, brown, or green; hard as steel file or harder; fracture shell-like or uneven; crystals equidimensional, often with 12 or 24 faces, also in grainy masses; several varieties (Figure 14–30).

FIGURE 14–19. Biotite mica, from Renfrew County, Ontario, showing a tabular, six-sided crystal and the capability of cleaving into thin sheets. Height of specimen is 3.1 inches (7.9 cm).

FIGURE 14–20. Scaly mass of chlorite, from Chester, Vermont. Width of specimen is 2.5 inches (6.3 cm).

FIGURE 14–21. Crystals of sphalerite, from Missouri, showing resinous luster. Width of specimen is 5.7 inches (14.5 cm).

FIGURE 14–22. Massive serpentine, from Chester County, Pennsylvania, showing waxy luster and lack of cleavage. Width of specimen is 2.8 inches (7.1 cm).

FIGURE 14–23. Fibrous serpentine, from Waldheim East Germany, showing delicate fibers. Height of specimen is 2.3 inches (5.8 cm).

FIGURE 14–25. Massive limonite, showing rounded surfaces. Width of specimen is 3.7 inches (9.4 cm).

FIGURE 14–24. Six-sided crystal prism of apatite, showing glassy to resinous luster. Height of specimen is 1.7 inches (4.3 cm).

FIGURE 14–26. Cleavage fragment of pyroxene, from Sydenham, Ontario, showing two directions of cleavage at about right angles as reflected by the top and right surfaces. Height of specimen is 1.1 inches (2.8 cm).

FIGURE 14–27. Cleavage fragment of amphibole, showing two directions of cleavage not at right angles. Width of specimen is 3.9 inches (9.9 cm).

FIGURE 14–28. Six-sided crystal prism of corundum, from Mineral County, Nevada. Height of specimen on the left is 1.5 inches (3.8 cm).

FIGURE 14–29. Granular mass of olivine, from Jackson County, North Carolina, showing glassy luster. Height of specimen is 2.9 inches (7.4 cm).

FIGURE 14–30. Twelve-faced crystal of garnet. Height of crystal is 2.2 inches (5.6 cm).

15 IGNEOUS ROCKS

In Chapter 14, I mentioned that rocks are made up of minerals or are aggregates of minerals. (Usually a few to several minerals constitute a rock but, in some cases, only one is present.) The study of rocks is **petrology.**

The three groups of rocks are igneous, sedimentary, and metamorphic. **Igneous rocks** (from the Latin word *ignis,* "fire") are literally "fire-formed" rocks: They result from the congealing of hot, melted rock deep within the earth—**magma**—or that extruded to the surface—**lava.** These rocks are the subject of this chapter. **Sedimentary rocks** (from the Latin *sedere,* "to settle") form from the accumulation of sediment—mineral crystals or particles of minerals and rocks or masses of organic matter—that ultimately solidifies into layered rock. **Metamorphic rocks** (from the Greek *meta,* "change" and *morphe,* "form") result from the alteration of other rocks at depth by heat, pressure, and chemically active fluids and gases. Sedimentary and metamorphic rocks are covered in Chapters 16 and 17.

Main Igneous Rock-forming Minerals

Only a few minerals—all silicates (Table 14–1)—make up the bulk of igneous rocks. They can be grouped into *light-colored* and *dark-colored* minerals. Mineral identification in rocks is more difficult because the mineral grains are small—a hand lens is usually required—and crystal form is less commonly developed. However, the basic physical properties still apply.

Light-colored minerals

The major light-colored minerals are the feldspars, quartz, and muscovite mica. Feldspar grains are recognized by their good cleavage in two directions at about right angles, common rectangular shape, and hardness greater than a knife. If the grains are pink, red, or flesh-colored, they can be considered orthoclase. If white, gray, or yellow, and possibly finely grooved, they can be called plagioclase. Quartz, commonly clear or white, although it may be almost any color, is glassy, lacks cleavage, and rarely shows crystal form. It appears to fill the spaces between other mineral grains. Muscovite, usually colorless, white, or yellow, is recognized by its breaking into thin, elastic flakes.

Dark-colored minerals

The dark-colored minerals—mostly dark green to black—include biotite mica, olivine, pyroxene, and amphibole. Not only dark, these minerals are heavier—have a higher specific gravity—than do the light-colored minerals because they contain iron and magnesium. Consequently, they are also called *ferromagnesian* (*ferro* means "iron") *minerals.* Biotite resembles muscovite but is darker. Olivine appears as green, glassy grains. Pyroxene—actually a group of minerals, as is amphibole—occurs as grains or short crystals with two directions of cleavage at right angles. Amphibole tends to occur in grains or elongate crystals that cleave in two directions not at right angles.

Classification and Identification of Igneous Rocks

Igneous rocks are classified on the basis of texture and mineral composition (Table 15–1). **Texture,** although strictly speaking meaning the size, shape and arrangement of mineral constituents of a rock, refers primarily to the *size* of mineral grains. It relates directly to the cooling history of the molten rock material from which an igneous rock originates. The main texture types are coarse-grained, fine-grained, glassy, porphyritic (not in Table 15–1), and fragmental.

In coarse-grained igneous rocks mineral crystals are large enough to be seen without a microscope—a hand lens may be needed—and most are of about the same size. Visualize a magma as a molten slush of liquid, gas (mostly water vapor and carbon dioxide) and suspended, early-formed crystals. If the magma cools slowly and uniformly, as it does at considerable depth beneath the earth's surface, ions have ample time to combine and form large, equal-sized crystals.

Fine-grained rocks possess crystals discernible only with a microscope. Such small crystals imply a much more rapid cooling of magma than do coarse-grained rocks, such as that which occurs near the surface or in the interior of lava flows.

Glassy rocks lack distinct crystals even when viewed with a microscope. They form under conditions of such rapid cooling that ions have been unable to combine into a crystal structure. Such rocks may form, for example, on the surfaces of lava flows or where lava enters water.

Rocks of the preceding textural groups may be "two-textured" or **porphyritic.** Two sizes of crystals are evident, implying two stages and rates of cooling: larger crystals from slower cooling, smaller crystals from faster cooling, both types in the same rock. If the larger crystals are particularly abundant, the rock is a **porphyry.**

Fragmental or *pyroclastic* ("fire-fragmental") rocks differ

Table 15-1. CLASSIFICATION OF IGNEOUS ROCKS

TEXTURE	MINERAL COMPOSITION: Light-colored minerals mostly		Dark minerals mostly	Dark minerals entirely
	Feldspar exceeds [1]B, A, P; quartz	Feldspar exceeds A, B, P; no quartz	P, O, A exceed feldspar; no quartz	O, P, A; no feldspar or quartz
Coarse-grained	GRANITE AND RELATED ROCKS	DIORITE	GABBRO	PERIDOTITE PYROXENITE HORNBLENDITE
Fine-grained	FELSITE (e.g., Rhyolite, Andesite)		BASALT	
Glassy	OBSIDIAN PUMICE		BASALT GLASS (Rare)	
Fragmental (=Pyroclastic)	TUFF VOLCANIC BRECCIA AGGLOMERATE			

[1]B=biotite, A=amphibole, P=Pyroxene, O=olivine; listed in order of relative abundance (most abundant first) in the sections above.

notably from the others. They consist chiefly of broken rock fragments—with glass—blown out during volcanic eruptions.

Now, let's examine the major igneous rocks, beginning with the coarse-grained ones.

Granite and related rocks

Granite (Figures 15-1 to 15-3) is a coarse-grained igneous rock composed mostly of orthoclase feldspar with quartz, and minor dark minerals, usually biotite or amphibole (Table 15-1). Muscovite may also occur.

Other granitic rocks are not readily distinguished in the field. *Syenite* resembles granite but lacks quartz; *quartz monzonite* contains about as much plagioclase as orthoclase; and in *granodiorite,* plagioclase exceeds orthoclase. Granitic rocks vary from white to dark gray to pink and red. **Pegmatite** is a very coarse-grained rock with crystals about an inch (a few centimeters) to many feet long. Most commonly of granitic composition (Figure 15-4), somewhat exotic minerals, such as topaz, tourmaline, and spodumene, may be also present. Pegmatite forms in a late stage of crystallization during which a gas-enriched, very fluid magma allows ions to move freely and form unusually large crystals.

Diorite

Generally darker than the granitic rocks—light gray to green—because of more dark minerals, **diorite** (Figure 15-5) consists chiefly of plagioclase, with amphibole and biotite, and commonly small amounts of pyroxene and orthoclase. Quartz is lacking or insignificant. Diorite is intermediate in composition between the

FIGURE 15–1. Granite from St. Cloud, Minnesota. Most of the lightest grains are of orthoclase feldspar; the light gray glassy grains—as in the center of the specimen—are of quartz; the dark grains are of amphibole and biotite. Width of specimen is 3.9 inches (9.9 cm).

FIGURE 15–2. Granite porphyry, implying two stages of molten-rock cooling. The long, light crystals are of orthoclase feldspar. The large crystal in the lower left is 0.9 inch (2.3 cm) long.

FIGURE 15–3. Exposure of weathered, jointed granite porphyry, showing characteristic rounded forms.

FIGURE 15–4. Granite pegmatite, of mostly feldspar (white), quartz (light gray), and muscovite (dark gray to nearly black). Height of the muscovite cluster in the upper right is 2.9 inches (7.4 cm).

granitic rocks and gabbro. It grades into granite through the intermediate granodiorite.

Gabbro

Darker still than diorite—dark gray, green or black—**gabbro** (Figure 15–6) commonly contains more dark minerals than plagioclase feldspar. The main dark minerals are pyroxene with olivine and minor amphibole. Quartz is absent.

Peridotite, pyroxenite, and hornblendite

These three rocks, generally green to black, are solely or nearly all of dark minerals. **Peridotite** (Figure 15–7) is mostly of olivine and pyroxene; in **pyroxenite,** pyroxene predominates; and **hornblendite** contains mostly amphibole. (It would seem logical to use the nonspecific *amphibolite* if uncertain that the amphibole present is hornblende. But *amphibolite* is reserved for a metamorphic rock.) Quartz is absent, and feldspar is almost always lacking.

Felsite

A fine-grained, light-colored igneous rock, **felsite** (Figures 15–8, 15–9) is a general term for the fine-grained equivalents of granitic rocks and diorite. Thin edges of flakes or chips of felsites tend to be translucent. If the texture is porphyritic and larger crystals are identifiable, you might attempt a more specific rock name. So, if considerable orthoclase and quartz are present—with perhaps biotite or amphibole—as in granite, the rock is **rhyolite.** Likewise, a light-colored rock with identifiable crystals chiefly of plagioclase, along with amphibole and biotite and other minerals found in diorite, is **andesite**—or porphyritic andesite or andesite porphyry (Figure 15–9) if larger crystals are particularly abundant.

Basalt

The fine-grained, dark-colored equivalent of gabbro is **basalt** (Figure 15–10). It appears dull, almost velvety. Thin edges of flakes or chips of this rock generally are not translucent as in felsite. Numerous, smooth cavities or *vesicles,* representing entrapped gas bubbles released as the lava cooled, are common in the upper part of basalt lava flows. These result in *vesicular* structure (Figure 15–10), found in other fine-grained and glassy rocks as well. The vesicles may later fill with mineral matter.

Glassy rocks

Obsidian (Figure 15–11) is a natural glass with shell-like fracture. Characteristically black, it is also gray, brown, or red. The edges of thin flakes are transparent or translucent. Native Americans fashioned knives, projectile points, and other implements and weapons from obsidian. Most obsidians have a composition similar to that of granitic rocks. **Pitchstone,** a variety of obsidian, owes its resinous or pitchlike luster to a relatively high water content. **Pumice** (Figure 15–12) is a glass froth, of glass fibers separated by numerous vesicles. Because it is so porous and the gas bubbles are sealed, pumice floats on water. It is white, gray, yellow, or brown. Like obsidian, pumice has a composition like that of granitic rocks.

Scoria is a cindery, slaglike rock with more vesicles than vesicular basalt—there may be as much empty space as rock—and with larger vesicles than pumice. Reddish brown to gray and black, it commonly is of basaltic composition. Brown to black **basalt glass** is relatively rare. This is because basaltic lavas are "thinner" or less viscous than felsitic lavas, and ions can combine more readily to form crystalline rather than noncrystalline rocks.

Fragmental rocks

As gases under pressure are released from a magma rising toward the earth's surface, fragments of new and old lava are blown out from volcanic vents. Fragments smaller than 0.2 inch (4 mm) constitute *volcanic ash* and *dust;* those larger than about 1.3 inches (32 mm) are called *bombs*—rounded, ejected while molten—and *blocks*—usually angular, ejected after solidification.

FIGURE 15-5. Diorite from Los Angeles County, California. The light grains are largely of plagioclase feldspar, the dark grains of amphibole and biotite. Width of specimen is 4.1 inches (10.4 cm).

FIGURE 15-7. Peridotite, from Hualalai Volcano, Hawaii, Hawaii. It consists entirely of dark minerals, chiefly olivine and pyroxene. Height of specimen is 3.5 inches (8.9 cm).

FIGURE 15-6. Gabbro, from Salem, Massachusetts. Dark minerals exceed plagioclase feldspar. Width of specimen is 4.5 inches (11.4 cm).

FIGURE 15-8. Felsite (rhyolite), from Castle Rock, Colorado, showing shell-like fracture. Width of specimen is 4.2 inches (10.7 cm).

FIGURE 15-9. Felsite (andesite) porphyry, from the Spanish Peaks region, Colorado, showing conspicuous, dark, amphibole crystals in a fine-grained matrix, implying two stages of molten-rock cooling. Height of specimen is 4.5 inches (11.4 cm).

FIGURE 15-10. Basalt, from near Naalehu, Hawaii, Hawaii, with numerous gas bubble cavities (vesicular structure). Width of specimen is 4.5 inches (11.4 cm).

FIGURE 15-11. Obsidian, showing excellent shell-like fracture. Width of specimen is 6.7 inches (17.0 cm).

FIGURE 15-12. Pumice, from Pumice Desert, Crater Lake, Oregon, showing glass fibers and gass bubble cavities. Width of specimen is 4.9 inches (12.4 cm).

FIGURE 15-13. Volcanic breccia, from Owens River, Nevada. This fragmental rock contains coarse, angular, volcanic rock fragments within a finer tuff matrix. Width of specimen is 5.8 inches (14.7 cm).

Tuff is a fine-grained, fragmental rock largely of volcanic ash and dust. It has a rough feel from the sharp ash and dust particles. Tuff is relatively lightweight and usually light-colored: white, gray, yellow, light brown, or pink.

Volcanic breccia (Figure 15–13)—mostly of large angular fragments—and **agglomerate**—mostly of large rounded fragments—are fragmental rocks of largely volcanic blocks and bombs mixed with ash and dust. They tend to be gray, yellow, brown, and red.

Occurrence of Igneous Rocks

Of the rocks exposed on the land surface, an average of 34 percent are crystalline—igneous and metamorphic—and an average of 66 percent are sedimentary. Crystalline rocks, however, underlie all sedimentary rocks. By volume, crystalline rocks constitute about 95 percent of the earth's crust to a depth of about nine miles (15 km). One-half of the crystalline rocks exposed are igneous.

Igneous rocks occur in two types of rock bodies: extrusive and intrusive. **Extrusive** bodies result from magma *extruded* on the surface, whereas **intrusive** bodies form as magma *intrudes* into fractures of surrounding rock at depth or melts and assimilates it (Figure 15–14). Percentages of extrusive and intrusive rocks exposed on the land surface are similar.

Extrusive rock bodies

Extrusive rock bodies are of two main types: lava flows and volcanoes. These are covered in Chapter 8. Besides basalt and felsite in flows, porphyries of these rock types also occur in these extrusive bodies. Volcanic glass commonly forms on the surfaces of

FIGURE 15–14. Block diagram showing occurrence of extrusive and intrusive igneous rock bodies.

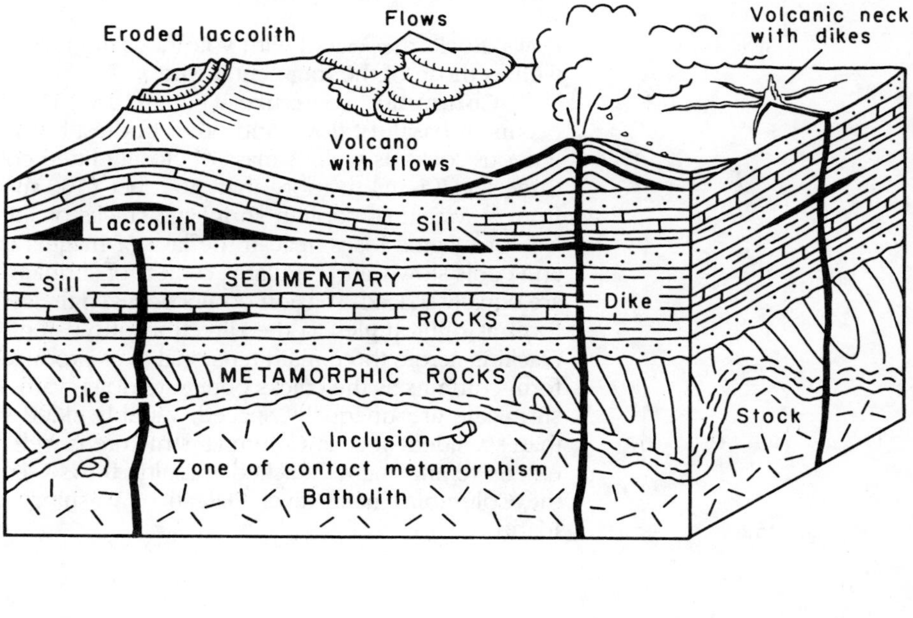

FIGURE 15–15. Columnar structure in an intrusive of felsite porphyry, Devils Tower, Devils Tower National Monument, northeastern Wyoming.

flows. Rocks associated with volcanoes are those of lava flows as well as those of the fragmental type.

Columnar structure (Figure 15–15) most commonly occurs in basalt of flows and other bodies, but is found in other igneous rocks as well. A mass of closely fitted columns less than three feet (1 m) to more than 300 feet (100 m) long, it results from a system of cooling shrinkage cracks or joints that form as lava or magma loses heat. If the lava or magma is homogeneous and cools slowly and regularly, centers of shrinkage or contraction are equally spaced. Tension occurs between the centers, and cracks form at right angles to the direction of tension. Ideally, double cracking across three main tensional axes produces hexagons that form columns as the cracks penetrate downward. If the centers of shrinkage are unequally spaced, variably sided columns—other than six-sided—occur. Columnar structure is well displayed in the extensive and many "stacked" basaltic flows (see Figure 8–7) of the Columbia–Snake River Plateau of Washington, Oregon, and Idaho.

FIGURE 15–16. Pegmatitic granite dike (light) intruded into quartz biotite schist. Width of the dike is 2.5 feet (0.8 m).

FIGURE 15–17. Light felsite dikes intruded into slate and schist and connected to light sills in the highest hill to the left. The tallest trees are about 40 feet (12 m) high. Homestake Mining Company open cut, Lead, Black Hills, southwestern South Dakota.

Intrusive rock bodies

Volcanic necks are cylindrical masses of rock sealing the vents and conduits of former and recently dormant volcanoes, readily visible upon removal of the more easily erodible surrounding rocks. They are common in northwestern New Mexico and adjacent Arizona. Rocks common in volcanic necks are coarse- and fine-grained porphyries, and glassy and fragmental rocks.

Dikes (Figures 15–16, 15–17) are tabular intrusive bodies—their length and breadth is great in relation to their thickness—and form as magma squeezes into vertical or near-vertical fractures. They are less than a few tenths of an inch to hundreds of feet thick and may extend for many miles. Dikes tend to occur in groups—following fracture systems—and frequently radiate from volcanic necks (Figure 15–14). Columnar structure develops in some dikes—forming at right angles to the cool fracture walls—the columns resembling stacked cordwood. Dikes resistant to erosion form rock walls that traverse the countryside. One region of abundant dikes is the Spanish Peaks area of southern Colorado.

Basically all fine-grained rocks and their porphyries occur in dikes, as well as coarse-grained porphyries, gabbro, peridotite, and pegmatite.

Sills (Figure 15–17) are also tabular intrusives that differ from dikes in having been forced *between* layers of older rock. They commonly occur where magma has been injected into weak, easily penetrated, stratified rock. Sills are of similar dimensions to dikes and may be offshoots of them. They may also display columnar structure; an example is in the basaltic, 1000-foot(300 m)-thick, Palisades sill exposed along the Hudson River in New Jersey. Sills differ from buried lava flows in that baked or altered rocks occur above and below them, fragments of the surrounding rock may occur in them, and they lack vesicular structure and weathered or eroded tops.

Rocks in sills are similar to those found in dikes except for pegmatite. Those of basaltic composition prevail.

Laccoliths are lens-shaped—in vertical section—intrusives, formed as relatively thick or viscous magma, fed from below or a side, forces its way between rock layers and domes up overlying strata. They have an arched top, consequently, and a rather flat base. Laccoliths are roughly circular or elliptical in plan view, up to thousands of feet thick and many miles in diameter. Well-known laccoliths occur in the Henry Mountains of southeastern Utah. Resistant laccoliths stand as prominent hills or mountains, frequently surrounded by concentric ridges of eroded, domed rock layers (Figure 15–18). Sills may be intimately associated with laccoliths—they differ only in being thinner and more extensive—as may dikes.

Most igneous rocks are found in laccoliths. Coarse-grained porphyritic rocks are particularly common.

Batholiths are the largest of intrusives and the largest rock bodies of the earth's crust. They may cover thousands of square miles and assume elongate, elliptical, or circular shapes. Batholiths

FIGURE 15–18. Eroded laccolith of felsite porphyry flanked by upturned edges of sedimentary strata. Bear Butte, near Sturgis, southwestern South Dakota.

tend to first increase in size with depth but may decrease in size with greater depth. Emplaced at depths of thousands of feet beneath the surface, these intrusives are exposed only where extensive erosion has occurred. Fragments of older surrounding rock—*inclusions* (Figure 15–14)—are often preserved within an emplaced batholith. Batholiths characteristically occur along the axes of mountain belts. Well-known examples in western North America are the Coast Range batholith of western British Columbia and the Idaho batholith of central Idaho and western Montana.

Granitic rocks are typically found in batholiths, but gabbro and diorite may also occur. Coarse-grained porphyries, too, can be expected in places.

Stocks are essentially small batholiths, less than about 40 square miles (100 square kilometers) in exposed area. Certain stocks are simply offshoots of batholiths. Both stocks and batholiths erode to topographically high areas, and may display well-developed jointing.

Rocks of stocks are similar to those of batholiths—in being generally granitic—but diorite, gabbro, and peridotite are also relatively common.

Selected Readings

CHESTERMAN, C. W. *The Audubon Society Field Guide to North American Rocks and Minerals.* New York: Alfred A. Knopf, Inc., 1978.

MOTTANA, A., R. CRESPI, and G. LIBORIO. *Guide to Rocks and Minerals.* New York: Simon and Schuster, Inc., 1978.

SORRELL, C. A. *Rocks and Minerals.* New York: Golden Press, 1973.

16 SEDIMENTARY ROCKS

Sedimentary rocks consist of solidified or lithified *sediment*—particles of minerals or rock fragments, mineral crystals, or organic matter. Sediments such as gravel, sand, and lime mud, for example, lithify to conglomerate, sandstone, and limestone. Sedimentary rocks of particles or fragments are **clastic** (from the Greek *klastos,* "broken into pieces"), those of mineral crystals or organic matter are **nonclastic.**

Exposed on 66 percent of the land surface, on the average, sedimentary rocks are more apt to be seen on the landscape than igneous and metamorphic rocks. In oceanic areas, too, sediments prevail over other earth materials.

Main Sedimentary Rock-forming Materials

Clastic rocks consist chiefly of quartz, feldspars, rock fragments, micas, clay minerals, iron oxides, and calcite (Table 16–1). The clay minerals are hydrated—water is bound with them chemically—aluminum silicates; kaolinite is an example. Individual clay minerals can be identified only with sophisticated laboratory equipment. The most common iron oxides are hematite and limonite; they serve to color and cement clastic sediments. Quartz and calcite constitute particles as well as cement sediments.

Nonclastic, chemically precipitated rocks are made up mostly of calcite, dolomite, gypsum, anhydrite, halite, quartz and related minerals, hematite, limonite, and apatite (Table 16–2). The carbon-bearing rocks consist of plant material.

Classification and Identification of Sedimentary Rocks

Clastic rocks are classified on the basis of the size, first, and the composition, second, of sedimentary particles. Composition of constituents is the sole basis for the general classification of nonclastic rocks.

Table 16-1. CLASSIFICATION OF CLASTIC SEDIMENTARY ROCKS

PARTICLE SIZE	*MAIN CONSTITUENT(S)*	*SEDIMENT*	*ROCK*
Coarse (>.08 inch or 2 mm)	Rock fragments, quartz	Gravel	Conglomerate (rounded particles) Breccia (angular particles)
Medium (0.002–0.08 inch or 0.06–2mm)	Quartz Feldspar Rock fragments	Sand	Sandstone Quartz Sandstone Feldspar Sandstone Rock Fragment Sandstone
Fine (<0.002 inch or 0.06 mm)	Clay minerals, quartz	Silt (coarser) + Clay (finer) = Mud	Mudstone (blocky) Siltstone Claystone Shale (fissile)
Coarse to Fine	Calcite	Limy gravel, sand, or mud	Clastic Limestone

Clastic rocks

Coarse-grained clastic rocks (Table 16–1) consist of lithified **gravel,** made up of an appreciable amount of particles greater than 0.08 inch (2 mm) in diameter. Ideally, "appreciable" means 50 percent or more, but geologists may call a sediment "gravel" with only 10 percent gravel-sized particles. Gravel includes *boulders* (greater than 10.1 inches or 256 mm), *cobbles* (2.5 to 10.1 inches or 64 to 256 mm), and *pebbles* (.08 to 2.5 inches or 2 to 64 mm).

Lithified gravel is **conglomerate** (Figure 16–1) if the particle edges are rounded, **breccia** (Figure 16–2) if the particle edges are relatively sharp or angular. (Loose sediment of coarse angular particles may be called *rubble.*) Besides the coarse-grained constituents or rock fragments—most common—or minerals, these rocks commonly have a *matrix* of finer sediment, a mineral *cement,* or both. A more detailed name, then, is based on the framework particles, matrix, and cement. How detailed this name is depends on you. A coarse-grained clastic rock of largely quartz pebbles in a sandstone matrix and cemented by calcite may be variously called: Possible names are pebble conglomerate; quartz-pebble conglomerate; sandy conglomerate; sandy, quartz-pebble conglomerate; or limy (or *calcareous*), sandy, quartz-pebble conglomerate. Coarse-grained clastic rocks grade into volcanic agglomerate and volcanic breccia as the amount of igneous fragmental rock (see Chapter 15) material increases.

Medium-grained clastic rocks consist of lithified **sand,** of particles 50 percent or more of which are 0.002 to 0.08 inch (0.06 to 2 mm) in diameter. Sand is obviously gritty when rubbed between the fingers. A **sandstone** (Figure 16–3) with conspicuous coarser or finer particles is a conglomeratic or muddy (described under fine-grained rocks) sandstone. Sandstones with 50 percent or more quartz, feldspar, or rock fragments may be termed **quartz**

FIGURE 16–1. Conglomerate, of rounded pebbles of chert and other rock fragments. Cavities represent places once occupied by pebbles. Height of specimen is 4.5 inches (11.4 cm).

FIGURE 16–2. Breccia, of angular, pebble-sizea chert fragments. Height of specimen is 3.3 inches (8.4 cm).

FIGURE 16–3. Sandstone, of largely quartz and feldspar, from the west shore of Tomales Bay, Marin County, California. Sand grains are visible in the right foreground. Width of specimen is 4.8 inches (12.2 cm).

FIGURE 16–4. Shale, from Somerville, New Jersey, showing the capability of breaking into thin sheets. Width of specimen is 4.7 inches (11.9 cm).

FIGURE 16–5. Siltstone, from near Newhall, California, showing blocky character, unlike the thin-sheet fracturing of shale (Figure 16–4). Width of specimen is 4.2 inches (10.7 cm).

sandstone, feldspar sandstone, or **rock fragment sandstone.** Other minerals, too, if conspicuous, may enter into the rock name; another common mineral, mica, gives rise to a mica or *micaceous* sandstone. Quartz is the most common mineral of sandstones and may constitute 90 percent or more of such rocks. Feldspar in appreciable amounts is uncommon; a sandstone with 25 or 30 percent or more feldspar may also be called *arkose.* Another name—in disfavor with some—is *graywacke,* which denotes a muddy (15 percent or more) sandstone of angular particles of quartz, feldspar, rock fragments, and mica, in varying proportions.

As with the coarse-grained rocks, sandstone names may be as complex as you wish to make them, with modifiers given in order of increasing abundance of the contained minerals. A sandstone consisting mainly of framework particles of quartz (40 percent), feldspar (20 percent), and rock fragments (15 percent), and held together by a muddy (15 percent) matrix and calcite cement (10 percent), could be called a limy (or calcareous), muddy, rock fragment, feldspar, quartz sandstone. Medium-grained clastic rocks grade into tuff as the amount of volcanic ash and dust (see Chapter 15) increases. A sandstone with considerable ash or dust is a tuff (or *tuffaceous*) sandstone.

Fine-grained clastic rocks are made up of lithified **silt** or **clay**—the two are collectively called **mud**—particles 50 percent or more of which are less than 0.002 inch (0.06 mm) in diameter. Silt particles are barely discernible with the unaided eye and slightly gritty between the fingers and teeth. (Yes, some geologists actually "chew" sediment to estimate the particle size!) Clay particles are indistinguishable with the unaided eye, completely lack grittiness between the teeth, and are smooth or slippery when rubbed between the fingers. *Clay,* then, refers to a sediment of *clay-sized* particles. It may consist, in part, of clay minerals but others, such as quartz and feldspar, may be present as well. A rock rich in clay minerals sticks readily to the tongue and exudes a strong earthy odor after being breathed upon.

Lithified mud is **mudstone** if *blocky*—fragments break out of exposures in small blocks—or **shale** (Figure 16–4) if *fissile*—fragments break out into thin sheets or plates parallel to the layering or bedding. Predominantly silt or clay material lithifies to **siltstone** (Figure 16–5) and **claystone;** these rocks are generally not fissile.

In the field, fine-grained rocks are named largely on the basis of color, which reflects composition. Red shale (or mudstone) usually results from the iron oxide hematite, finely divided. Limonite tends to produce yellow or brown shales. Green shale commonly results from iron compounds, but here the iron ion has a charge of +2 instead of +3, and is said to be "ferrous," not "ferric." Green may be produced, too, by other minerals; another source is glauconite—a silicate of iron, magnesium, aluminum, and potassium. Grey or black shales commonly contain variable amounts of organic matter: the darker the shale, the more the organic matter. (Shales rich in organic matter also tend to be more

FIGURE 16–6. Fossil-bearing or fossiliferous limestone, from near Guttenberg, Iowa, made up largely of shells of brachiopods (see Figure 18–6). Width of specimen is 5.1 inches (12.9 cm).

FIGURE 16–7. Porous travertine, a variety of limestone, deposited by springs (see Figure 6–8); Yellowstone National Park, Wyoming. Height of specimen is 5.8 inches (14.7 cm).

FIGURE 16–8. Alabaster, a fine-grained variety of rock gypsum, from Rapid City, South Dakota. Height of specimen is 4.3 inches (10.9 cm).

FIGURE 16–9. Chert, with abundant oölites, from Centre County, Pennsylvania. Width of specimen is 4.0 inches (10.2 cm).

FIGURE 16–10. Flint, a variety of chert, showing shell-like fracture; Dover Cliffs, England. Width of specimen is 4.5 inches (11.4 cm).

fissile.) But shales may be black from finely divided iron sulfide. Remember that the streak—the color of the powdered mineral (see Chapter 14)—of the iron sulfide pyrite is greenish black.

Fine-grained rocks may be named also on the basis of coarser admixtures or strictly on the basis of composition. Appreciable amounts of gravel or sand result in conglomeratic mudstone or sandy shale. Considerable carbonate or mica gives rise to limy (or calcareous) or micaceous shale.

Nonclastic rocks

Carbonate rocks contain 50 percent or more carbonate minerals, chiefly calcite—forming **limestone**—or dolomite—forming **dolostone.** (Some also use the term *dolomite* for the rock.) Both limestone and dolostone are easily scratched with a knife and come in an array of colors from white to black. Limestone bubbles readily in dilute acid, dolostone weakly or not at all. If the drop of acid is placed on the powdered dolostone grooved out by the knife, however, the bubbling is vigorous.

Limestone is more common than dolostone and may consist of several constituents: organic remains, complete or broken, such as fossil shells or coral (see Chapter 18); carbonate rock particles; oölites or pisolites; and calcite cement. A limestone of largely unbroken fossils is an organic or fossiliferous limestone (Figure 16–6). *Chalk*—visualized readily in the form of blackboard chalk—is a crumbly organic limestone consisting largely of the limy parts of one-celled plants and animals. A rock of broken fossils or carbonate rock fragments is a **clastic limestone** (Table 16–1) and has all the characteristics of any other clastic rock. *Coquina* (from the Spanish word for "shellfish") is a clastic limestone of coarse-grained fossil fragments. *Lithographic limestone* is the lithified equivalent of lime mud, and was once used extensively for lithography because of its fine, even grain. *Oölites* (Figure 16–9)—less than 0.08 inch (2 mm) in diameter—and the larger *pisolites* are spheroidal or ellipsoidal particles formed by chemical precipitation in shallow, wave-agitated waters, such as that on the Bahama Banks. Limestones with such particles predominating are oölitic or pisolitic limestones. A minor limestone precipitated from spring water—hot or cold—is *travertine* (Figure 16–7). It includes the dripstone and flowstone of caves (see Chapter 6).

Limestones grade into dolostones, and both rocks grade into noncarbonate clastic rocks. Intermediates include dolomitic limestone, calcitic dolostone, muddy limestone (dolostone), and sandy limestone (dolostone).

Evaporites—the most common of which are **rock gypsum, rock anhydrite,** and **rock salt** (Table 16–2)—form by evaporation of brines. That is, these salts precipitate from brines that are concentrated by evaporation. A necessary requirement for evaporation is an arid climate. In addition, briny waters must be partially separated from the main ocean so the salt concentration can increase by evaporation. For example, gypsum and anhydrite are precipitating today at the margins of the Persian Gulf. Evaporites also form in certain lakes in arid regions.

Table 16-2. CLASSIFICATION OF NONCLASTIC SEDIMENTARY ROCKS

MAIN CONSTITUENT(S)	*ROCK GROUP*	*ROCK*
Calcite	Carbonate	Limestone
Dolomite		Dolostone
Gypsum	Evaporite	Rock Gypsum
Anhydrite		Rock Anhydrite
Halite		Rock Salt
Quartz and related minerals (chalcedony, opal)	Siliceous	Chert (Light) Flint (Dark) Diatomite
Hematite, limonite, siderite	Iron-rich	Ironstone (Massive)
Magnetite, hematite, pyrite		Iron Formation (Banded)
Plant material	Carbonaceous	Peat Lignite Coal Bituminous Coal Anthracite Coal
Apatite	Phosphate	Phosphate Rock (Phosphorite)

All of the three common evaporites are crystalline—as the igneous rocks—coarse- to fine-grained, and most commonly white, although they may be variously tinted. Each consists almost entirely of gypsum, anhydrite, or halite, with few mineral impurities. All can be readily scratched with the knife, rock gypsum with the fingernail. Rock salt, of course, has a salty taste. *Alabaster* (Figure 16–8) is massive, fine-grained, white or tinted rock gypsum that is shaped into ornamental objects.

All three evaporites are commonly associated with one another and frequently associated with carbonate rocks—especially dolostone—and shales. Rock anhydrite readily alters to rock gypsum in exposures by the addition of water. Because it is easily dissolved, rock salt occurs at the surface only in very arid regions, such as at Great Salt Lake, Utah and Death Valley, California.

Siliceous rocks consist of silica—silicon + oxygen (SiO_2)—in the form of quartz, including microcrystalline chalcedony, and related minerals, especially noncrystalline opal. **Chert,** the most common siliceous rock, is dense, tough, harder than a knife, and exhibits shell-like fracture. It may include oölites (Figure 16–9) if silica has replaced oölitic limestone. **Flint** (Figure 16–10) is dark gray to black chert that owes its color to included organic matter. Its toughness and hardness have favored its use in the fashioning of aboriginal weapons and implements. *Jasper* is yellow, brown, or red chert colored by the iron-bearing limonite or hematite.

Chert (and flint) occurs in nodules and beds. Such nodules (defined under Sedimentary Rock Structures) are discrete, often irregular—even lumpy—masses in carbonate rocks. Bedded chert is associated with shales and iron formations (described under iron-rich rocks). Chert may form by the direct, inorganic precipitation of a jellylike silica mass; replacement of other rocks, such as limestones; or organic precipitation of silica by such organisms as diatoms (mentioned directly), radiolarians (single-celled animals with siliceous shells), and sponges—certain ones secrete spicules or needlelike elements of opal.

Other siliceous rocks include usually white or gray *siliceous sinter,* deposited by hot springs or geysers (geyserite) (see Figure 6–7), and **diatomite** (Figure 16–11), composed of the microscopic siliceous shells of single-celled plants, *diatoms.* Diatomite, or the more crumbly **diatomaceous earth,** is usually white, chalklike—but does not bubble in acid—is somewhat gritty when rubbed between the fingers, and generally lacks the earthy odor of clay minerals when breathed upon. Diatomites form in both fresh and marine waters.

Iron-rich rocks of sedimentary origin contain about 15 percent or more iron, which is, unfortunately, difficult if not impossible to estimate. A little iron goes a long way, and rocks usually appear to contain more iron than they actually do.

The iron-rich rocks can be placed into two groups: ironstone and banded iron formation. **Ironstone** (Figure 16–12) may be of almost any sedimentary rock type, but especially of mudstone, sandstone, and limestone; it is massive to poorly banded; and it is nearly always of Cambrian age and younger. The most common minerals in ironstone are hematite, limonite, and siderite. Siderite, a carbonate mineral, looks like brown calcite, but is heavier and somewhat harder—though it can still, however, be scratched by the knife; it often alters to limonite. Much of the hematite ironstone consists of oölitic hematite or hematite that has replaced fossils or filled spaces between them. The Silurian Clinton Formation, extending from New York to Alabama, contains such hematite ironstone, a commercial iron ore. **Banded iron formation** (Figure 16–13), frequently called *taconite,* is thinly bedded or banded, usually with magnetite, hematite, or pyrite (uncommonly) alternating with chert (jasper) or quartz. It is nearly always of Precambrian age. The extensive banded iron formation near western Lake Superior is renowned.

Iron-rich rocks have been formed largely in shallow seas, either by direct precipitation of iron minerals or by the later iron-replacement of other rocks. Minor iron-rich rocks (ironstones) have formed also in lakes or bogs (bog iron ore).

Carbonaceous rocks—bearing carbon—consist of peat (although strictly this is a sediment rather than a rock) and coal. **Peat** (Figure 16–14) is a brown-to-black accumulation of plant matter that closely resembles compressed or chewing tobacco, depending on its depth in a layer. Under pressure, peat converts to **coal**—a series of lignite, bituminous, and anthracite, in order

FIGURE 16-11. Chalklike diatomite, from Santa Barbara County, California, showing thin layering. Height of specimen is 1.1 inch (2.8 cm).

FIGURE 16-14. Peat, from Cambridge, Massachusetts, showing plant fragments on a surface of layering. Height of specimen is 2.9 inches (7.4 cm).

FIGURE 16-12. Ironstone, of oölitic hematite, from the Clinton Formation (Silurian) of Clinton, New York. Height of specimen is 3.1 inches (7.9 cm).

FIGURE 16-15. Lignite coal, from southwestern North Dakota, showing cracking and crumbling upon exposure to air. Width of specimen is 4.3 inches (10.9 cm).

FIGURE 16-13. Banded iron formation, of largely alternate bands or layers of hematite (lustrous) and jasper (iron-bearing chert; dull). Width of specimen is 7.7 inches (19.6 cm).

FIGURE 16-16. Bituminous coal, from Pittsburgh, Pennsylvania, showing pitchy luster and banding. Width of left front surface is 3.1 inches (7.9 cm).

of increasing rank and subjection to increased pressure and heat. As carbonaceous material passes through the coalification progression, it increases in luster, carbon content, and heating value, and decreases in moisture and content of gases. **Lignite** (Figure 16–15) is brown to black, lacks luster, commonly displays the original wood structure, and burns readily with a smoky flame and strong odor. It may contain as much as 40 percent moisture, and slacks or crumbles readily upon exposure to the atmosphere. **Bituminous coal** (Figure 16–16) differs from lignite in being strictly black, having a glassy or pitchy luster evidenced in distinct bands or layers, lacking original wood structure, and not slackening. **Anthracite coal** (Figure 16–17) displays glassy to submetallic luster and shell-like fracture, and burns initially with difficulty and with little smoke or odor.

Coal originates within a humid climate where luxurious vegetation proliferates in swamps. Plant matter only partially decomposes in the standing water and accumulates to form peat. Burial by overlying sediment compresses the peat, and coalification occurs. Older coals tend to be of higher rank because of the likelihood of deeper burial, but exceptions occur. Coals of higher rank also occur near intrusive igneous rock bodies and where rocks have been intensely folded. In western Pennsylvania, for example, where strata are essentially flat-lying, the coal is bituminous; in eastern Pennsylvania, anthracite is mined from enclosing folded strata. Most coals are of Pennsylvanian, Cretaceous, or Tertiary age. They are associated primarily with shales, mudstones, and sandstones.

Phosphate rock or **phosphorite** consists largely of apatite, but the mineral is unrecognizable in the rock in the field. It resembles limestone—but does not bubble when acid is applied—and is heavier. Phosphate rock is commonly black, but may be almost any color from white to black. It commonly contains oölites, pisolites, sand-sized pellets and larger nodules, and fossil bone, fish teeth and scales, and shells. Interbedded strata are frequently limestone, mudstone, and chert. Much phosphate rock occurs, for example, in the Permian Phosphoria Formation of Idaho and adjacent states. Phosphate rock has originated largely in the sea by direct precipitation or later replacement of limestones. Upwelling and the consequent bringing up of nutrient-rich water toward the surface seems conducive to phosphate precipitation.

Sedimentary Rock Structures

Sedimentary rock structures are usually those features of a larger scale that are best seen in rock exposures. They help identify sedimentary rocks and aid in interpreting the environment in which a sediment was laid down. Certain structures were formed at the time a sediment was deposited, others originated later.

Stratification—or layering or bedding—is the most distinctive structure of sedimentary rocks, formed when a sediment is laid down but emphasized by exposure to weathering and erosion (Figure 16–18). Certain igneous rocks, such as basalt in flows,

FIGURE 16–17. Anthracite coal, from Jeddo, Pennsylvania, showing glassy luster and shell-like fracture. Height of specimen is 2.3 inches (5.8 cm).

are stratified as well, but the feature is best developed in sedimentary rocks. Bedding is normally horizontal, but **cross-bedding** (see Figure 3–4), occurring at some angle, is produced when water or wind currents deposit sediment on a slope. (Imagine the inclined beds inside a sand dune, viewed as if the dune were sliced through by a huge knife.) Other structures formed when a sediment is deposited are such bottom surface features as ripple marks, mud cracks, and the pockmarking raindrop impressions. **Ripple marks** (Figure 16–19) are parallel ridges and swales effected by currents and waves moving sediment. **Mud cracks** form as wet, fine-grained sediment dries out. Both mud cracks and raindrop impressions indicate a sediment surface has been exposed to the air. Fossils (see Chapter 18) may be considered organic sedimentary structures, whether they are actual remains of organisms or such traces as tracks, trails, or borings.

Other sedimentary structures are formed largely after sediment deposition by chemical action. Nodules and concretions are notable examples; both are segregations of mineral matter. **Nodules** are considered here as bodies of mineral matter unlike that of the host rock in which they are found. Chert and flint nodules in carbonate rocks are examples. **Concretions** (Figure 16–20)—not consistently defined—are treated here as discrete bodies—commonly spheroidal, ellipsoidal, or disk-shaped—with a composition similar to that of the host rock except for the mineral cement, which is commonly calcite, silica, or iron oxide. Spheroidal types are frequently called "cannonball" concretions. Concretions may have a fossil as a nucleus and a concentric internal structure.

FIGURE 16–18. Stratification or layering of sedimentary rocks, emphasized by uneven erosion of hard and soft strata. The deeper, inner gorge is cut into metamorphic and intrusive igneous rocks. Grand Canyon National Park, northwestern Arizona. (Photograph 238-197-63 by the National Park Service.)

FIGURE 16–19. Ripple marks, in quartz sandstone, from Missouri. Length of specimen front to back, is 11.4 inches (29.0 cm).

FIGURE 16–20. Concretions of siltstone, from near Medora, North Dakota. Width of left, broken concretion is 3.3 inches (8.4 cm).

Origin of Sedimentary Rocks

Sediments ultimately originate at the earth's surface by the in-place breakup of rocks by chemical and physical means—weathering (see Chapter 23). The sediments are transported by water, wind, and ice and finally laid down in a myriad of environments: streams, deltas, dunes, beaches, and shallow seas, to name a few. They accumulate, layer upon layer, and finally lithify to rock. The lithification process involves a reduction in the pore space between particles or crystals, by compaction or cementation. In **compaction**—most significant in the fine-grained sediments—pore space is reduced by the pressure resulting from the weight of overlying rock and sediment. Particles become closely packed and so lithified. In **cementation,** spaces between particles are filled with mineral matter—particles are cemented together—and lithification ensues by a distinctly different means. Mineral matter can be precipitated into pore space by either fresh water—ground water—or by sea water.

Selected Readings

CHESTERMAN, C. W. *The Audubon Society Field Guide to North American Rocks and Minerals.* New York: Alfred A. Knopf, Inc., 1978.

MOTTANA, A., R. CRESPI, and G. LIBORIO. *Guide to Rocks and Minerals.* New York: Simon and Schuster, Inc., 1978.

SORRELL, C. A. *Rocks and Minerals.* New York: Golden Press, 1973.

17 METAMORPHIC ROCKS

In Chapter 15, I mentioned that metamorphic rocks originate from the alteration of pre-existing rocks by the action of heat, pressure, and chemically active fluids and gases. Such rocks are exposed on the land surface as extensively as the igneous rocks, and most of the rocks beneath sedimentary rocks on continents are metamorphic. So metamorphic rocks deserve a closer look although they, at times, are more difficult to identify than igneous or sedimentary rocks.

Main Metamorphic Rock-Forming Minerals

Since metamorphic rocks derive from igneous and sedimentary rocks, we might suspect that metamorphic rock–forming minerals may be similar to those of these two rock groups. Common, therefore, are the minerals quartz, feldspars, micas, amphiboles, pyroxenes, and carbonates. Others, however, such as graphite (see Figure 14–1), talc (see Figure 14–14), chlorite (see Figure 14–20), serpentine (see Figures 14–22, 14–23), and garnet (see Figure 14–30), are characteristic of—but not restricted to—metamorphic rocks. Still others, such as staurolite (see Figure 17–4) and kyanite, are restricted to them.

Classification and Identification of Metamorphic Rocks

Classifying metamorphic rocks (Table 17–1) is based primarily on whether they are foliated and secondarily on grain size and composition. **Foliation** is a layering in metamorphic rocks caused by the parallel arrangement of platy or elongate minerals. Micas are platy minerals, and amphiboles, for example, tend to occur as elongate crystals. Pressure-induced foliation gives rise to **rock cleavage,** a tendency for rocks to split into sheets along well-defined surfaces. Don't confuse this with mineral cleavage, the tendency for *minerals* to split along planes because of weaknesses in internal atomic bonding.

Table 17-1. CLASSIFICATION OF METAMORPHIC ROCKS

FOLIATED		*NONFOLIATED*
Mostly from Regional Metamorphism	*From Mechanical Metamorphism*	*Mostly from Contact Metamorphism*
Slate Schist Gneiss	Fault Breccia Mylonite	Metaconglomerate and Metabreccia Quartzite Marble Hornfels Serpentinite

Foliated rocks

Slate (Figure 17–1) is a fine-grained foliated rock—the grains are invisible with the unaided eye or hand lens—that splits readily into thin sheets and is said to have good slaty cleavage. When slate is struck with a metal object, such as the handle of a pocketknife, you often hear a tinkling sound. Slate is usually gray or black but may be green, purple, red, brown, or yellow. It develops from the metamorphism of shale, tuff, and other fine-grained rocks. *Phyllite* differs from slate in displaying coarser-grained, visible mica flakes that impart a satiny sheen on rock cleavage surfaces.

Schist is a medium- to coarse-grained foliated rock in which the mineral grains—particularly the micaceous ones—are readily visible. Rock cleavage is distinct (Figure 17–2), but not as well developed as in slate and phyllite. The numerous varieties of schist—resulting in a spectrum of colors—are named after prominent minerals. Examples include talc schist, graphite schist, chlorite schist, biotite schist, garnet mica schist (Figure 17–3), and staurolite mica schist (Figure 17–4). Schist derives from a wide group of rocks including shale, sandstone, tuff, basalt, felsite, and gabbro.

Gneiss is a coarse-grained metamorphic rock with poor foliation and rock cleavage. Minerals—more quartz and feldspar than micaceous minerals—tend to be segregated into light and dark bands that are frequently folded or contorted in exposures. Most gneiss has the composition of granite—granite gneiss—but composition varies widely and, therefore, so does color. Varieties of gneiss are named after rock composition (as for granite gneiss) or major minerals: Examples include mica gneiss, biotite gneiss (Figure 17–5), and hornblende gneiss. Gneiss derives from a host of rocks, including granitic rocks, diorite, gabbro, felsite, tuff, shale, and sandstone.

Fault breccia and **mylonite**—relatively minor metamorphic rocks—form in fault zones by the mechanical mashing and pulverizing of rock against rock. Fault or crush breccia closely resembles sedimentary breccia and must be recovered from a fault zone to substantiate its identity. It forms near the surface under relatively low confining pressure—that equal in all directions. Mylonite (from the Green *mylon,* "a mill") is a fine-grained, flint-

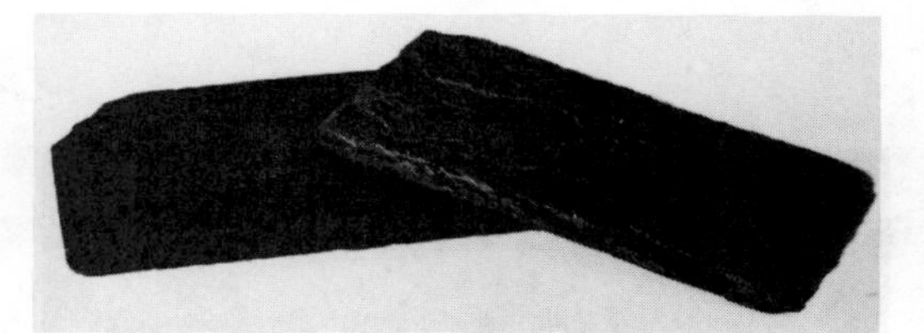

FIGURE 17–1. Black slate, showing good slaty rock cleavage. Length of the upper (right) slab is 7.0 inches (17.8 cm).

FIGURE 17–2. Exposure of quartz biotite schist showing distinct, steeply dipping rock cleavage. The weathered, upturned edges of the schist are bent to the right because of creep, the slow, downslope movement of rock, soil, and sediment (see Chapter 7).

FIGURE 17–3. Garnet mica schist—looking along the direction of well-developed foliation—from New York, showing two well-developed garnet crystals. Width of the rock specimen is 4.2 inches (10.7 cm).

FIGURE 17–4. Staurolite mica schist, from Little Falls, Minnesota, showing foliation surfaces and intergrown (crossed) staurolite crystals in the upper right. Width of the rock specimen is 4.8 inches (12.2 cm).

FIGURE 17–5. Biotite gneiss, from Uxbridge, Massachusetts, showing light bands of quartz and feldspar and dark bands of mostly biotite. Width of the specimen is 4.1 inches (10.4 cm).

FIGURE 17–6. Metaconglomerate, of pebbles of quartz and quartzite within a quartz biotite gneiss. Length of the largest pebble in the upper right is 5.1 inches (12.9 cm).

FIGURE 17–7. Pink marble, from Tate, Georgia, showing a sparkle of recrystallized crystal faces and cleavage surfaces. Width of the specimen is 4.1 inches (10.4 cm).

like or felsitelike, tough rock that often appears streaked as if smeared out. It has, literally, passed through a metamorphic mill. Depth and confining pressure are greater than for fault breccia. Both fault breccia and mylonite, obviously, vary considerably in composition, depending on the rocks involved in faulting.

Nonfoliated rocks

Metaconglomerate (Figure 17–6) and **metabreccia** are nonfoliated metamorphic rocks differing from their sedimentary equivalents in being generally tougher and having fractures passing *through* particles instead of around them. The same characteristics are present in **quartzite,** metamorphosed sandstone.

Marble (Figure 17–7) is metamorphosed limestone and dolostone, and sparkles readily in reflected light. This is because of recrystallization (discussed under Origin of Metamorphic Rocks) during the metamorphic process, resulting in interlocking crystals commonly coarser than the grains of the original carbonate rock. Marble, then, consists mostly of calcite and dolomite—easy scratching by the knife and bubbling in acid identify it—but various accessory minerals, such as garnet, feldspar, graphite, and apatite, might be present. Marble is normally light-colored, but almost any color from white to black is possible. A "marbling" effect from impurities striping or blotching the rock is common.

Hornfels is a fine-grained, dense, nonfoliated rock that resembles dark chert, basalt, or nonfoliated slate. It is usually gray or black. Hornfels usually derives from shale, but basalt and felsite are among other possible parent rocks.

Serpentinite, as the name implies, is a rock composed mostly of serpentine and having the general properties of that mineral. Normally green, it varies from yellow to black. Common associated minerals are olivine, pyroxene, and amphibole. It serves as a host rock for nickel and chromium ores. Serpentinite derives from the alteration of igneous rocks, such as peridotite and pyroxenite, and metamorphic rocks, exemplified by amphibole schist.

Origin of Metamorphic Rocks

Through metamorphism, a rock changes in texture, composition, or internal structure as it adjusts or responds to a new geologic environment. Such change—an attempt to reach an equilibrium with new environmental conditions—occurs within the solid (nonmolten or nonliquid) state, apart from the domain of igneous activity. New minerals form, but no real change occurs in the bulk composition of the rocks so affected—elements are not added or removed from the system but merely reshuffled.

Agents of metamorphism

Agents responsible for metamorphic changes are increased *heat* and *pressure* and *chemically active fluids and gases.* They tend to work in concert more often than not.

Increased heat increases the activity of ions and causes the breakdown of minerals; but increased heat causes, too, more vigorous chemical reactions and the recombination of ions to form new minerals in a new thermal environment. More heat drives off water from minerals and provides more water in the pore space between grains. The increased heat-water combination softens rocks, causing them to yield more readily under pressure. The heat derives from (1) that released continually from the earth's core or from the decay of radioactive minerals; (2) molten rock or magma; and (3) frictional heat generated as rock masses grind against one another during large-scale earth movements.

High pressure caused by the weight of a pile of rock strata makes underlying rocks more compact as pore space between grains is reduced. But this downward pressure doesn't contribute to metamorphism. It is sideways-directed pressure at considerable depth—by large-scale earth movements—that causes a directional property in many metamorphic rocks. Large crystals, pebbles, and fossils may be smeared out to several times their length. But more important is the development of foliation as newly formed platy or elongate minerals align themselves at right angles or parallel to the directed pressure. We've already examined in Chapter 12 how extensive vertically foliated rocks form from earth-plate squeezing during mountain building. In the second case, shearing pressures are involved: two opposing pressures gliding past one another. Visualize this process by imagining thin plastic disks cut out with a hole puncher—mimicking mica flakes—placed in a ball of modeling clay. As the ball of clay is sheared or smeared between two boards—representing opposing shearing pressures—the plastic disks tend to become parallel, simulating foliation.

Chemically active fluids and gases act as solvents, speed up chemical reactions, and serve as media by which ions can be transported through pore space between grains to form new minerals. The most important fluid is water, derived from (1) original sea water trapped in sediments; (2) magmas; and (3) water molecules combined chemically in minerals.

A significant process often involving the three metamorphic agents in combination is **recrystallization**—a change in crystal size or mineral composition as material is dissolved and removed in some places and added elsewhere. For example, pressure-solution occurs at points of greater stress and addition or precipitation—growth of minerals—occurs in places of lesser stress. Increased heat accelerates solution and precipitation, and fluids and gases enhance solution and facilitate the recombination of ions for precipitation.

Kinds of metamorphism

Contact metamorphism develops from a body of magma in *contact* with surrounding rocks. Increased heat, primarily, causes rock alteration, although pressure, too, is influential. A baked zone of contact metamorphism (see Figure 15–14) surrounds an exposed intrusive mass, commonly a few hundred feet wide and only rarely a few thousand feet wide. The degree or intensity of metamorphism becomes gradually less away from the intrusive body, reflected particularly by the presence of key minerals; those that form, however, depend largely on the rock types intruded and the temperatures reached. These minerals—and associated rocks—can be perceived as occurring in concentric cylinders that gradually envelop the intrusive body. Staurolite, for example, may occur adjacent to the intrusive, followed progressively outward by garnet, biotite, and chlorite.

The nonfoliated rocks form mostly by contact metamorphism (Table 17–1). Rare is the rock, however, that forms strictly by this process.

Regional metamorphism, or large-scale metamorphism not linked intimately with obvious intrusive bodies so as to produce distinct, extensive baked zones, has formed the greatest volume of metamorphic rock. This type of metamorphism involves areas of thousands of square miles—specifically, elongate regions that have undergone mountain building (see Chapter 12). Tremendous pressures, primarily, have caused the rock alteration, although high temperatures have contributed significantly—both occurring at considerable depth, presumably in the roots of mountain belts. Again, as for contact metamorphism, key minerals reflect the degree or intensity of metamorphism. Staurolite schist (Figure 17–4), for example, has formed at higher pressures and temperatures than garnet schist (Figure 17–3).

Plate tectonic theory explains that huge earth plates have slammed together, squeezing sediments and rocks, viselike, between them. As rocks of a descending plate reach a critical, hot depth they melt, along with certain of the deepest squeezed metamorphic rocks. Lighter magma from the melting rises and intrudes the metamorphic rocks, producing particularly the extensive batholiths at the cores of mountain belts. When we see extensive tracts of metamorphic rocks and associated intrusives—even where planed relatively flat, as in the great Canadian Shield of eastern and central Canada—we can conclude they represent the roots of former mountain belts, brought to view by much erosion over long periods of time.

The foliated rocks originate mostly by regional metamorphism (Table 17–1). Slate, schist, and gneiss form a metamorphic series, with gneiss produced by the greatest pressure and accompanying heat. In progressing from slate to gneiss, foliation and rock cleavage become poorer and grain size increases.

Mechanical or **dynamic metamorphism** occurs largely by rock deformation. The type producing fault breccia and mylonite merges gradually into regional metamorphism with greater depths and the accompanying higher pressures and temperatures. Another rather intriguing type—*shock metamorphism*—occurs by the impact of meteors or underground nuclear explosions. In a fraction of a second shock waves produce very high pressures and temperatures. In the outer part of the zone of shock metamorphism surrounding a meteor crater, the rock is shattered. Toward the site of impact, natural glass filling cracks in fractured rocks increases so that mostly glass occurs at the center of the crater. On the moon and similar bodies with abundant meteoritic impact craters, shock metamorphism must be a significant near-surface metamorphic process.

Selected Readings

CHESTERMAN, C. W. *The Audubon Society Field Guide to North American Rocks and Minerals.* New York: Alfred A. Knopf, Inc., 1978.

MOTTANA, A., R. CRESPI, and G. LIBORIO. *Guide to Rocks and Minerals.* New York: Simon and Schuster, Inc., 1978.

SORRELL, C. A. *Rocks and Minerals.* New York: Golden Press, 1973.

18 FOSSILS

Fossils and Fossilization

Fossils (from the Latin *fossilis,* "dug up") are any evidence of past life—plant or animal—preserved by natural means or causes in materials of the earth's crust. The word "fossil" usually conjures up something like a bone or shell—a *body fossil.* But indirect evidence, such as a track, trail, or burrow (Figure 18–1)—the so-called *trace fossil*—is within the fossil realm as well. Fossils normally are entombed in sediment or sedimentary rock. But they are also stuck in asphalt and the resin of cone-bearing trees, and frozen in ice. Most fossils are millions of years old, and purists believe organic remains or traces must have been buried prior to the beginning of written history to be called fossils. Others do not strictly adhere to this age technicality. And how can you always know how old a supposed fossil is? But then, a line must be drawn somewhere—or must it? Human remains in cemeteries today are not considered fossils, but it isn't always clear in other contexts when such remains should be called "fossils." Having been buried might be a more significant criterion. After all, a fossil is something "dug up," as stipulated by the literal meaning of the word. The study of fossils is **paleontology.**

Let's consider a scenario from an organism's death, to its burial and preservation—*fossilization*—to discovery of the fossil. A clam living at the edge of a sea will do nicely to begin this mental journey. The clam dies when a snail bores a hole through its shell and feasts on the insides. The shell opens, gets knocked around by waves and currents for a time, but soon is covered by sand, protected at least from further destruction by organisms. Encroachment of the sea over the land begins, and layer upon layer of sediment piles up over the clam. Thousands of years pass, then millions. The weight of the overlying sediment compacts the sand with the clam, and other sediment, to rock. One-half of the calcite shell is dissolved by acid-bearing ground water seeping slowly through the porous sandstone, leaving only an impression. But the other half of the shell remains because it is in a muddy layer that prevents the ground water from moving through. Earth movements raise the land, and a deep valley is carved through the thick sequence of sedimentary rocks, exposing the clam-bearing sandstone.

One hot summer afternoon, a tired backpacker ambles along the river, his mind occupied mostly with his heavy pack and aching shoulders. He nearly topples over as a boot hooks on the sharp broken edge of a sandstone block. Turning back, his eye catches sight of a white shell in the reddish rock. Kneeling down to examine it closely, he notices a small, neat hole in one part of the shell; the other part is strangely gone, only an impression remains. He looks up, traces the sandstone layer from which the block has fallen—and wonders, his aching shoulders forgotten.

Our hypothetical clamshell could have done worse. It could have been distorted by recrystallization (see Chapter 17) or mis-shapen or even destroyed by metamorphism of the rock containing it. Other possible changes include replacement of the shell by, say, silica or pyrite or the simple filling of a shell cavity in the rock by the same materials.

We might infer from the clam story that *hard parts* and *rapid burial* of an organism are conducive to fossil preservation and later discovery. Both hinder decay and destruction. Rapid burial occurs most readily in the watery environments, particularly the sea, lakes, ponds, and bogs. Soft parts, however, are occasionally preserved, as in the case of the elephantine mammoths frozen in Siberia and leaves remaining behind as films of carbon.

Oldest Fossils

The oldest known fossils fall into two groups, both of which probably go equally unnoticed by most people. One group includes microscopic spheroids and rods believed to be bacteria and algae (single-celled plants). The other is the **stromatolites,** easily visible, thinly layered, usually limy, domal structures resembling, internally, sliced-through cabbage heads. Algae have probably largely produced these structures. The oldest fossils in both groups are dated at about 3.5 billion years and were recovered in South Africa and Western Australia.

Earliest life must have originated prior to 3.5 billion years ago because the oldest known fossils likely do not record the earliest life. And life must have populated the earth after about 4.5 billion years ago (see Chapter 11), generally considered by geologists as the time of the earth's origin.

Classifying, Identifying, and Naming Fossils

Classifying fossils, or any objects for that matter, means arranging them into a previously devised system, into progressively more specific categories. Specifically, it is usually the assigning of a *species*—the basic biologic unit—or of a *genus* to a higher, more inclusive category. Classification of man—the species—illustrates this (Table 18–1). Such a classification allows us to *portray rela-*

Table 18-1. GENERAL CLASSIFICATION OF MAN.

Kingdom ANIMALIA: Organisms that require complex organic food and are capable of movement voluntarily.

Phylum CHORDATA: *Animals* with an upper, lengthwise support—the *notochord*—, an upper nerve cord and brain, and gill arches and pouches in the embryonic stage.

Class MAMMALIA: *Chordates* with hair and mammary glands for suckling the young.

Order PRIMATES: *Mammals* with four generalized limbs, each with five fingers or toes bearing nails.

Family HOMINIDAE: *Primates* with manlike features, and a larger and more functional brain than the apes and other closely related primates.

Genus HOMO: Hominids unlike other genera in the family.

Species: *Homo sapiens.*

tionships among similar animals (in this case) and to *organize* considerable specific information. *Identifying* fossils means placing *individual specimens* into pre-established groups. Paleontologists are most often concerned with identification, but classification, at least, is always implied. Preliminary identification only is an aim of this chapter; for more specific identification, consult the readings at the end of this chapter, the numerous others available, or, if you are particularly serious, a paleontologist at a university or museum.

Why is each fossil—or living organism—given a separate, often long, and seemingly unpronounceable "scientific" name? Well, to begin with, scientists everywhere must be able to communicate internationally, and a single, *Latinized* name for each organism provides this. Regardless of the scientist's language, the scientific name always appears the same. As for the so-called "common" names, many organisms don't have them, and an organism may have several, depending on peoples' whims or the locale. Consider, for example, a bush with edible fruit, *Amelanchier alnifolia,* that has at least four common names: juneberry, serviceberry, sarviceberry, and saskatoon. The scientific name tends to eliminate such possible confusion. Granted, the Latinized names are often long and may, at times, be awkward to pronounce. But the presumed awkwardness frequently dissipates with familiarity and frequent use. Do the names *Hippopotamus, Rhinoceros, Chrysanthemum,* and *Delphinium* seem unduly difficult? In these cases the scientific and common names of the two animals and two plants are the same. One final point: Scientific names often tell something about an organism. *Rhinoceros* is formed of the Greek *rhin-,* "nose" and *keras,* "horn": an animal with one or two horns on its "nose," or better, snout.

I'll not dwell on scientific names in this chapter. They are best learned through constant use and total immersion, which

FIGURE 18–1. Trace fossil, a presumed burrow filling (*Ophiomorpha*) of a marine ghost shrimp, showing the characteristic knobby surface; Cretaceous; near Rhame, North Dakota. The burrow may branch vertically or horizontally. Length of specimen at the base is 5.7 inches (14.5 cm).

FIGURE 18–2. Impressions of sponges of Silurian (*left;* saucerlike *Astraeospongia* from Decatur County, Tennessee) and Devonian (*right;* knobby *Hydnoceras* from near Alfred, New York) age. Length of *Hydnoceras* on the right is 5.3 inches (13.5 cm).

FIGURE 18–3. Solitary, hornlike corals of Devonian age, showing radial partitions. At the top is *Heliophyllum,* at the bottom is *Zaphrentis* (length of this specimen is 3.4 inches or 8.6 cm).

FIGURE 18–4. Colonial coral (*Lithostrotionella*) of many-sided prisms, showing radial and crosswise partitions; Mississippian; Keokuk, Iowa. Width of the specimen on the right is 3.1 inches (7.9 cm).

you'll have if you become seriously interested in fossils. Such names appear here mostly in the figure captions.

Main Fossil Groups

Fossils can be grouped into three categories: invertebrates, vertebrates, and plants. **Invertebrates** are "animals without backbones," those that lack vertebrae or a segmented spinal column, such as the clam, snail, or lobster. **Vertebrates,** having vertebrae, include the fishes, amphibians, reptiles, birds, and mammals. Man, of course, is a vertebrate; this would have been obvious if "Subphylum Vertebrata" had been included under "Phylum Chordata" in Table 18–1.

About 250,000 species or kinds of fossils are presently known, three-fourths of which are invertebrates. New species are discovered continually, and some geologists believe perhaps several million will ultimately be known. For comparison, about 1.5 million living organisms are known, most of which are invertebrates, chiefly insects.

Because of the great number of fossils, only a few can be mentioned in this chapter and on a larger group basis. Since invertebrates are most abundant and most frequently found as fossils, more space is allocated to them. By the way, if you have forgotten the geologic time terms, you might wish to slip a bookmark in Chapter 11 for easy referral. You will see them frequently from here on.

Invertebrates

We will examine only those invertebrates that are readily seen with the unaided eye and most apt to be observed by the traveler. **Protozoans**—single-celled animals—for example, are omitted because they are mostly microscopic, although they number as fossils in the tens of thousands.

Sponges (Precambrian to present) are evidenced in rocks most often by microscopic, limy or siliceous, needlelike elements, but occasionally as globular, cylindrical, vaselike, or saucerlike impressions (Figure 18–2). Most sponges are marine today and presumably were so in the past.

Corals (Ordovician to present) have generally limy skeletons that, from Paleozoic rocks, are usually hornlike (Figure 18–3) or resemble many-sided prisms (Figure 18–4) or cylinders, frequently grown together to form a tight, rounded mass. Both growth forms are usually partitioned lengthwise or radially as well as crosswise. Many-tentacled animals—like sea anemones—sat at the ends of the horns, prisms, or cylinders and secreted them. Living corals, and those from younger rocks, have assumed branchlike forms, forms resembling convoluted brains, and others. Today, corals are exclusively marine. The grown-together or colonial types form reefs in warm seas, and similar types must have done so in the past.

Bryozoans (Ordovician to present), pinhead-sized animals with tentacles and more complex than corals, secrete mostly limy colonies that are encrusting, stemlike or branching-twiglike, and

FIGURE 18–5. Bryozoans. *Clockwise,* from lower right, are *Fenestella* (Devonian; White Mound, Oklahoma), *Archimedes* (Mississippian; Logan County, Kentucky), *Dekayella* (Ordovician; Cincinnati, Ohio), and *Hallopora* (Ordovician). *Archimedes* is 0.9 inch (2.3 cm) long.

FIGURE 18–6. Brachiopods. *Clockwise,* from upper right, are *Oleneothyris* (Paleocene; New Egypt, New Jersey), *Dictyoclostus* (Pennsylvanian; Wichita, Kansas), *Platystrophia* (Ordovician), and *Mucrospirifer* (Devonian; near Sylvania, Ohio). *Oleneothyris* is 1.9 inches (4.8 cm) long.

FIGURE 18–7. Snails. *Clockwise,* from right, are *Turritella* (Eocene; near Ariton, Alabama), *Volutospina* (Eocene; near Jackson, Alabama), *Fasciolaria* (Pliocene; near Clewiston, Florida), *Campeloma* (Paleocene; McKenzie County, North Dakota), and *Oliva* (Miocene; near Magnolia, North Carolina). All are marine except *Campeloma,* which is fresh water. Length of *Turritella* is 3.4 inches (8.6 cm).

FIGURE 18–8. Clams, *Crassatellites* (*left;* Cretaceous; near Enville, Tennessee) and *Pecten* (*right;* Miocene; Jones Wharf, Maryland). *Pecten* is 3.1 inches (7.9 cm) high.

lacy (Figure 18–5). The stemlike or branching-twiglike—or stony—bryozoans, found mostly in Ordovician and Silurian rocks, may have bumps or ridges covering the colonies. *Archimedes*—the screwlike fossil about which a lacy colony is spirally wrapped—characterizes Mississippian rocks in central North America. It was named after the Greek Archimedes, who invented the water screw for raising water. Bryozoans are mostly marine today and must have been so in the past.

Brachiopods (Cambrian to present) generally have limy shells of two parts that differ in size and shape (Figure 18–6). This implies that a plane of symmetry—an imaginary plane dividing the shell as equally as possible—would pass *across* the two parts and not between them, along the midline. The shells are variably shaped and may be smooth or ornamented with ridges, grooves, spines, and growth lines. Most brachiopods are marine today and must have been so in the past; more species, however, occurred in the past than now, and most are collected from Paleozoic rocks.

Of the *mollusks,* soft-bodied invertebrates characteristically with a shell and more complex than brachiopods, those most frequently found as fossils are snails, clams, and cephalopods. **Snails** (Figure 18–7) (Cambrian to present) usually have a limy, coiled shell that is not partitioned. They may be smooth or ornamented similarly to the brachiopods. Most live today in shallow seas, but many live in fresh water as well as on the land. The same was probably true in the past. Many freshwater shells tend to be thinner than marine shells and lack ornamentation.

Clams (Figure 18–8) (Cambrian to present) have limy shells of two parts somewhat similar to those of most brachiopods, but the plane of symmetry generally passes *between* the two shell parts. So the two parts are generally of similar size and shape. Where they are not, clams might be distinguished inside from brachiopods by rather large, shallow, oval or circular depressions that represent attachment scars of muscles that, upon contraction, close the two parts of the shell. Clams are most frequent in Mesozoic and Cenozoic rocks, whereas brachiopods are most abundant in Paleozoic rocks. Clams today prevail in shallow seas but they are common in *brackish* water—less salty than sea water but more so than fresh water—and fresh water as well. This pattern, presumably, is an extension from the past. Oysters, for example, are generally characteristic of brackish water, as found in bays and estuaries.

Cephalopod mollusks (Cambrian to present) today include relatively few species of squids, cuttlefish, the chambered nautilus, octopuses, and related groups, usually with poorly developed or no shells. But those of the past had well-developed limy shells, straight or coiled, external or internal, with partitions (Figure 18–9). External-shelled *nautiloid* cephalopods (Cambrian to present) possessed straight or slightly curved partitions—seen where the shell is stripped away—and *ammonoid* cephalopods (Devonian to Cretaceous) had wrinkled partitions. The chambered nautilus is the only living nautiloid. The extinct *belemnites* (Mississippian to Eocene), related to squids, had cigar-shaped internal shells. All cephalopods are marine today, and most likely were so in the past.

Arthropods, such as the crab, lobster, shrimp, and insects, are the most numerous of invertebrates today—especially the insects—but not all groups have left a good fossil record. The **trilobites** (Figure 18–10) (Cambrian to Permian), however, have. Segmented, like all arthropods, they are also tri- or three-lobed lengthwise, with a central lobe flanked by two side lobes. Although occurring throughout the Paleozoic, most trilobites are found in Cambrian, Ordovician, Silurian, and Devonian rocks. Geologists believe trilobites were probably all marine from their association with other marine fossils.

Three groups of *echinoderms,* or "spiny-skinned" animals, such as the living starfish, sea cucumbers, and sea urchins, are common as fossils: blastoids, crinoids, and sea urchins. **Blastoids** (Figure 18–11) (Silurian to Permian) have budlike skeletons of limy, fused plates with distinct, radial, depressed areas bisected

FIGURE 18–9. Cephalopods. *Clockwise,* from upper right, are two ammonoids, a belemnite, and a nautiloid: *Scaphites* (Cretaceous; near Linton, North Dakota), *Baculites* (Cretaceous; near Belle Fourche, South Dakota), *Belemnitella* (Cretaceous; near St. Georges, Delaware), and *Michelinoceras* (Devonian; Cayuga Lake, New York). Length of *Belemnitella* (extreme left) is 3.8 inches (9.6 cm). *Baculites* (lower right) and *Michelinoceras* (upper left) show the distinctive wrinkled and relatively straight partitions of ammonoids and nautiloids.

FIGURE 18–10. Trilobites, *Peronopsis* (*left;* Cambrian; Jince, Bohemia) and *Paedumias* (*right;* Cambrian; near York, Pennsylvania). Specimen on the left is 1.8 inches (4.6 cm) long.

FIGURE 18–11. Blastoids, two species of *Pentremites,* from Mississippian rocks at Monroe County, Illinois. Specimen at the left, 0.7 inch (1.8 cm) long, shows the point of attachment of a segmented stalk on the right.

FIGURE 18–12. Crinoids on bedding surface of limestone (Mississippian; LeGrand, Iowa). At least eight crowns—cups-with-arms—as well as several segmented stalks, are visible. Length of the largest crown in the lower right—excluding the stalk—is 2.9 inches (7.4 cm).

FIGURE 18–13. Reconstructed crinoid "garden" on a Mississippian sea floor. (Courtesy Field Museum of Natural History, Chicago; photograph GEO-80871.)

by food-gathering grooves. Most were attached to segmented stalks or columns that were anchored to the sea bottom.

Crinoids (Figure 18–12) (Ordovician to present), also known by the misleading term "sea lilies," have cuplike skeletons of more numerous plates than do blastoids, and they are arranged in circlets. Branched, armlike appendages extend upward. Most fossil crinoids were stalked, as were the blastoids. Disklike or star-shaped stalk segments are found more frequently than the crowns (cups-with-arms). Both crinoids and blastoids are particularly abundant in Mississippian rocks. Paleontologists imagine crinoid "gardens" (Figure 18–13) on limy, Mississippian sea floors, the numerous individuals gathering food with their outspread, armlike structures and swaying to and fro on flexible stalks in the waves and currents. Upon preservation their remains are seen scattered on bedding surfaces of limestone (Figure 18–12).

Sea urchins (Ordovician to present) have globelike, heart-shaped, or disklike limy skeletons, and lack stalks and armlike structures. "Regular" urchins show fivefold, radial symmetry, emphasized by top-to-bottom–trending bands of numerous plates. "Irregular" urchins (see Figure 21–4), including heart urchins and sand dollars, display bilateral symmetry. Petal-like impressions on the upper surfaces, through which breathing structures extend, characterize many of these. Sea urchins, and all other echinoderms today, are marine; presumably earlier ones were as well.

Graptolites (Figure 18–14) (Cambrian to Mississippian) appear as long, dark, carbon films in fine-grained clastic or carbonate rocks. They resemble narrow, saw-blade impressions with "teeth" on one or both sides. The "saw-blades" or branches occur singly, in groups, or in a netlike arrangement. "Teeth" on the branches are actually tubes or cups in which presumably lived tiny animals that, together, formed colonies. The colonies either floated in Paleozoic seas or were attached to sea bottoms. Graptolites, most often found in Ordovician and Silurian rocks, may be related to a group close to the vertebrates.

Vertebrates

Rare as fossils, vertebrates are most often evidenced as scattered teeth, skeletal bones, bony plates, spines, and scales. Study of these remains is detailed and complex, and it is difficult to place them, at times, in even one of the five major groups (Table 18–2). Fish remains are most frequent in Mesozoic and Cenozoic rocks, but very rarely are complete skeletons found (Figure 18–15). Several Paleozoic fishes were protected by bony plates. Sharks and their allies, with skeletons mostly of cartilage, are represented as fossils largely by teeth (Figure 18–16). Bony fish have left behind numerous scales, more abundant than their bones; thick, rhomboid scales covered the primitive types—such as the present-day gar—and thin, circular scales prevailed in the remainder.

Amphibians reached their heyday during the Late Paleozoic. They are believed to have evolved from one of the fleshy-finned fishes in the Devonian.

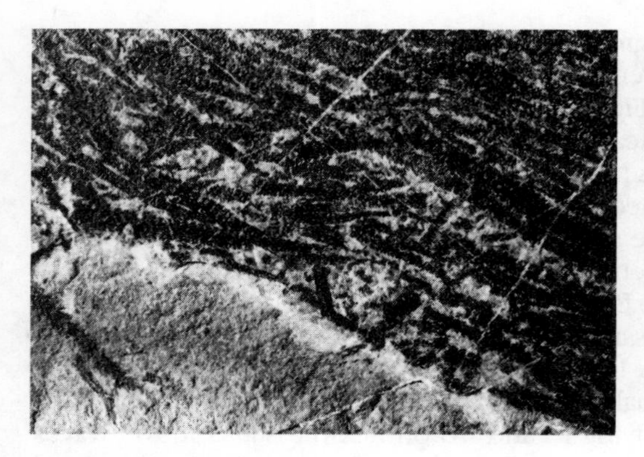

FIGURE 18–14. Graptolites, in limestone (Ordovician; near Overbrook, Oklahoma). Largest, isolated branch in the lower left is 0.6 (1.5 cm) long.

FIGURE 18–15. Bony fish (*Priscacara*) from Eocene rocks at Green River, Wyoming. Length of the specimen is 5.1 inches (12.9 cm).

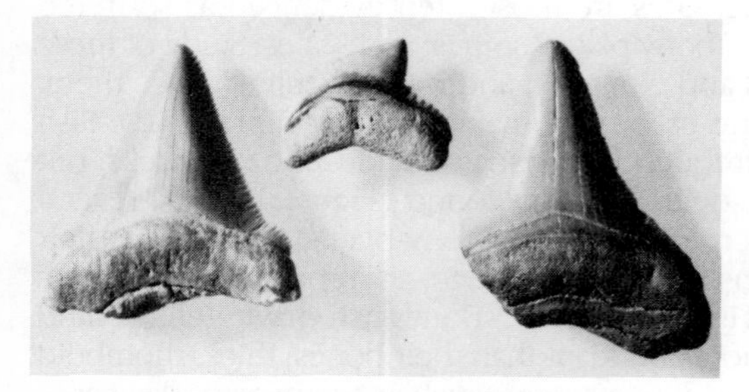

FIGURE 18–16. Shark teeth from middle Tertiary rocks. Height of the tooth on the right is 1.6 inches (4.1 cm).

FIGURE 18–17. Under view of a reconstruction of the long-necked, long-tailed, Jurassic dinosaur *Brontosaurus.* Person is 5.5 feet (1.7 m) high. Calgary Natural History Park, Calgary, southern Alberta, Canada.

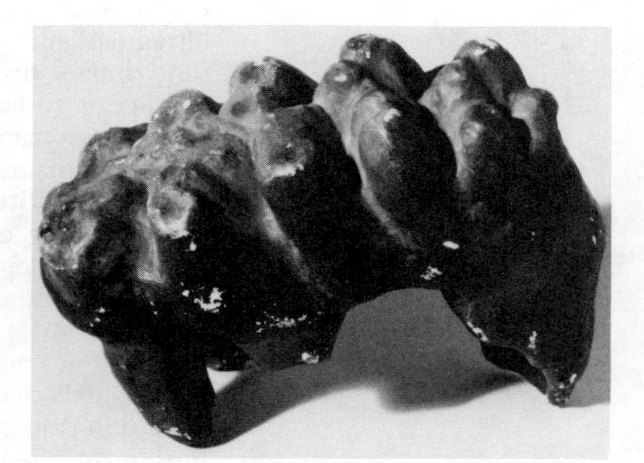

FIGURE 18–18. Plaster cast of a Pleistocene mastodon tooth, from Big Bone Lick, Kentucky, showing the relatively high points or cusps on the upper surface, here considerably worn. Length of the tooth is 6.7 inches (17.0 cm).

Table 18-2. SELECTED CHARACTERISTICS FOR DISTINGUISHING THE MAJOR GROUPS OF FOSSIL VERTEBRATES

CHARACTERISTIC	*FISHES (CAMBRIAN)**	*AMPHIBIANS (DEVONIAN)*	*REPTILES (PENNSYLVANIAN)*	*BIRDS (JURASSIC)*	*MAMMALS (TRIASSIC)*
1. Ball-and-socket joint at back of skull	Single	Single in early forms, double in later forms	Mostly single	Single	Double
2. Shallow grooves or pits on side of skull	Yes	Early forms only	No	No	No
3. Tooth type	Alike	Alike	Alike	—	Unlike
4. Roots of rear teeth	Single	Single	Single	—	Multiple
5. Scales	Mostly	Rare or lacking	Yes	No	No

*Age is earliest known occurrence; also shown on Figure 11-1.

Reptiles occur mostly in Mesozoic rocks: marine, freshwater, and terrestrial. Skeletal remains of certain marine types, such as the dolphinlike ichthyosaur, might be confused with huge fish. Bone fragments of the larger land reptiles—dinosaurs—are relatively common in places. Reptiles most likely evolved from an amphibian ancestor in the Pennsylvanian.

Birds are the rarest of vertebrate fossils; their characteristic, relatively light, hollow bones and skulls with large eye sockets occur mostly in Cenozoic rocks. Reptilian characters, such as teeth and backbone extending into the tail, of the earliest Jurassic bird imply a reptile ancestor for this group.

Mammal fossils are found mostly in Cenozoic rocks. Besides the characters given in Table 18-2, others aid in separating them from fossil reptiles: (1) a single opening in the skull for the nostrils as compared to two for most reptiles; (2) a bony roof of the mouth, lacking in reptiles except for certain groups, such as the crocodiles; and (3) each half of the lower jaw a single bone, compared to multiple bones in reptiles. These and other differences only partly facilitate the indistinct separation of gradational, latest Paleozoic mammal-like reptiles from reptiles and mammals. Such a problem, however, seems to clearly indicate that mammals arose from reptiles.

From the meager evidence, vertebrate paleontologists flesh out and reconstruct these animals (Figure 18-17), and speculate on their behavior and diet. For example, from the teeth of mastodons (Figure 18-18) and mammoths—with high points or cusps in mastodons and a flat grinding surface of folded enamel in mammoths—it seems clear to reason that they were leaf-chomping browsers, on the one hand, and grass-grinding grazers on the other.

Plants

Plants occur more commonly as fossils than vertebrates; those most frequently found are algae, lycopods, arthrophytes, ferns and seed-ferns, cycads and cycadeoids, ginkgos, conifers, and flowering plants.

Algae (Precambrian to present), the simplest of green plants, lacking true roots, stems, and leaves, are typified by the scumlike masses in ponds and by the seaweeds. As fossils, they are commonly preserved as stromatolites (mentioned under Oldest Fossils) and similar thinly layered masses. Diatomite (see Chapter 16) consists of the siliceous-secreting algae, diatoms.

Lycopods (Devonian to present), with true roots, stems, and leaves, are represented today by such small, inconspicuous plants as the club mosses or ground pines. But in the late Paleozoic they assumed tree size and contributed significantly to the vegetation of coal-forming swamps. These "scale trees" have rhomboid or circular leaf attachment scars on the trunks that resemble scales.

FIGURE 18–19. Impressions of leaflets of a frond of a seed-fern (*Pecopteris*) in a split concretion (Pennsylvanian; Mazon Creek, Illinois). Height of the concretion on the right is 5.2 inches (13.2 cm).

FIGURE 18–20. Impression of a ginkgo (*Ginkgo*) (right) of Paleocene age (near Almont, North Dakota) compared to a leaf of the sole living species (*Ginkgo biloba*). Consider the similarity in spite of a time separation of 60 million years! Length of the fossil leaf plus leafstalk is 3.3 inches (8.4 cm).

FIGURE 18–21. Petrified tree trunks representing a fossil forest buried by volcanic breccia, ash, and dust about 50 million years ago. Most of the trunks are of hardwoods. Specimen Ridge, Yellowstone National Park, northwestern Wyoming. Height of the person is 5.5 feet (1.7 m).

Arthrophytes (Devonian to present) get their name from the Greek *arthron,* "joint" and *phyton,* "plant." Who hasn't sat on a stream bank or railroad embankment and pulled apart the finely ribbed stems of horsetails or scouring rushes at their joints? Arthrophytes, with circlets of leaves and branches at the stem-joints, grew treelike in the late Paleozoic coal-forming swamps.

Ferns (Devonian to present) and **seed-ferns** (Figure 18-19) (Mississippian to Cretaceous) are similar in their usual feathery leaves of many leaflets; but ferns reproduce by spores and seed-ferns did so by pollen and seeds. Tree-sized representatives of both groups grew in late Paleozoic coal-forming swamps.

Cycads (Triassic to present) and **cycadeoids** (Permian to Cretaceous) are similar in their palmlike leaves and keglike to cylindrical trunks with rhomboid leaf attachment scars. But cycads reproduce by conelike structures and cycadeoids developed seeds from flowerlike structures embedded in trunks that frequently resemble large pineapple fruits. Both groups occur most commonly in Mesozoic rocks.

Ginkgos (Permian to present) (Figure 18–20) have fan-shaped leaves, divided or not, with veins radiating out from the point of attachment to the leafstalks. Prominent during the Mesozoic, only a single species lives today.

Conifers (Devonian to present), cone-bearing plants such as the pines, firs, and spruces, probably reached their maximum numbers during the Mesozoic, but are significant plants now as well. Primitive conifers—with straplike leaves—and scale trees, arthrophytes, and ferns and seed-ferns were the five main coal-forming plant groups of the late Paleozoic.

Flowering plants (Cretaceous to present) have dominated plant life on the earth since the late Cretaceous. They include two major groups: *monocots,* with one seed leaf and parallel-veined leaves—such as the grasses and lilies; and *dicots,* with two seed leaves and net-veined leaves—such as most deciduous trees and other broad-leafed plants. Most leaves and petrified wood (Figure 18–21) from late Cretaceous and younger rocks are of dicot flowering plants.

Uses of Fossils

Perhaps the most apparent use of fossils is to date, or determine the age of, rocks that contain them. For example, I already mentioned that the bryozoan *Archimedes* is an index of Mississippian age in central North America. Identification to genus or species is normally necessary for specific age determination, but knowing general groups, their age ranges, and their times of greatest abundance allows general age assignment. Trilobites are Paleozoic—most are Devonian and older—and graptolites usually signify an Ordovician or Silurian age for a rock. In related fashion, fossils help determine the equivalence or correlation (see Chapter 11) of rock layers in widely separated places.

Fossils may be so abundant as to form rocks·exclusively, or

nearly so, of their hard parts, analogous to rock-forming minerals. Coquina, chalk, diatomite, and coal are examples.

Together with sedimentary rocks, fossils aid in deciphering environments of the past (see Chapter 12). It may be difficult at first to accept that seas have covered North America several times since the beginning of the Cambrian. But fossil-bearing rock layers, stacked one above the other, tell us this is so. You may recall that corals, brachiopods, cephalopods, blastoids, crinoids, graptolites, and trilobites very likely lived in saline (or brackish) water during their existence, and are generally good marine indicators. Certain snails (Figure 18–7) and clams, however, serve to distinguish past fresh-water environments. In related fashion, fossils are good indicators of past climates. Silurian coral reefs in the presently temperate east-central United States, for example, signify previous tropical conditions there.

Fossils, too, document the progression of life through time, which generally has gone from simple to complex and from less varied to more varied. The order of coverage of fossil groups in this chapter—of the invertebrates, vertebrates, and plants—is toward more complexity for each group. To most paleontologists and biologists, the progression of life is explained by the concept of **evolution:** that later organisms develop or evolve from earlier organisms as their continually changing genetic make-up interacts with changing environments, and a continual selecting of the better goes on. In marked contrast is the concept of **creationism,** that all organisms were simply created by an omnipotent Creator. Granting that the fossil record clearly documents the **extinction**—dying out—of several organisms at different times and the later appearance of new organisms, this may imply the creation of organisms several times. I'll leave you to ponder the validity of both concepts. But keep in mind the fossil record of certain "intermediates," such as the mammal-like reptiles and the oldest fossil birds—with feather impressions to be sure, but also with reptilelike teeth.

Selected Readings

MACFALL, R. P., and JAY WOLLIN. *Fossils for Amateurs, a Guide to Collecting and Preparing Invertebrate Fossils.* New York: Van Nostrand Reinhold Company, 1972.

RANSOM, J. E. *Fossils in America.* New York: Harper and Row, Publishers, Inc., 1964.

RHODES, F. H. T., H. S. ZIM, and P. R. SHAFFER. *Fossils, a Guide to Prehistoric Life.* New York: Golden Press, 1962.

THOMPSON, I. *The Audubon Society Field Guide to North American Fossils.* New York: Alfred A. Knopf, Inc., 1983.

TIDWELL, W. S. *Common Fossil Plants of Western North America.* Provo, Utah: Brigham Young University Press, 1975.

How to do geology

19 THE GEOLOGIST'S APPROACH

Basic Premise: Uniformitarianism (Actualism)

Most geologists, in their attempt to unscramble geological puzzles, adhere to a basic premise: The processes shaping the earth and determining events today have also done so in the past. Or, simply, present-day causes can be used to explain the past. This premise, championed particularly by the Scottish geologist James Hutton (1726–1797), assumes that natural physical laws have remained unchanged, invariant through time. That processes and physical laws have been continuous and *uniform* through time has led naturally to the use of the term *uniformitarianism* for the premise. But a uniformity of causes does not imply also uniformity of conditions, results, or rates. Why should the rate of erosion on a mountain slope or the rate of sediment buildup in an ocean basin today, for example, necessarily equal that of a million years ago? Remember, the present is but a geological instant. It can hardly encompass all events of the past. The extensive cold climate and glaciers of 18,000 years ago are no longer with us. The last Ice Age is but one instance of different events occurring on the earth during the past and now.

A better term to some is *actualism*. Processes *actually* operating, or those inferred to be operating, can explain features and events of the past. The term *actualism* is less apt to mislead; it does not imply uniformity of anything other than processes and physical laws, as may the term *uniformitarianism*.

By either designation, uniformitarianism or actualism, geology can hardly be accomplished without its basic premise: that "the present is the key to the past."

How Do Geologists Think?

Geologists are scientists, and one may expect them to think creatively like other scientists. But all scientists do *not* think alike. So how do geologists differ?

Anne Roe, in *The Making of a Scientist*, analyzed the thinking processes of three groups of research scientists: biologists, physicists (both experimental and theoretical), and social scientists (psychologists and anthropologists). Four types of creative thinking emerged: thinking in terms of pictures or symbols ("visual imagery")—favored by biologists and experimental physicists; thinking in terms of (unspoken) words ("auditory-verbal")—used especially by theoretical physicists and social scientists; just knowing by a feeling of relationships ("imageless thought")—used mostly by experimental physicists and social scientists; and thinking through feelings of muscular tension ("kinesthetic thinking")—mentioned only by the social scientists.

Because most geologists deal or have dealt with natural objects in the field—like most biologists—we might infer that geologists commonly think in terms of pictures or symbols. A nearly universal behavior of geologists is the constant use of maps, cross sections (see Chapter 20), diagrams, and the like to help convey their ideas—either in publications or in their discussions. This results from exposure to a multitude of graphic aids during their training. Sophisticated graphics, in particular, are utilized to portray earth features, such as folds and fractures in rocks, three-dimensionally. These allow, quite naturally, for three-dimensional thinking. Who can always say when illustrations are conveying ideas or giving rise to their birth?

How to Mentally Attack a Geological Problem

Imagine flying over a broad stream valley in the United States Southwest. Below, you spot what appear to be broad, flat-topped benches flanking the valley on both sides—and at least at three levels. Are they lava flows through which the stream has cut? Man-made features? Do they relate directly to the stream in some way? You make a mental note of their location.

Weeks later, you drive through the valley by car. Examining road cut exposures, you discover the steplike benches consist chiefly of sand and gravel. Well, that rules out lava flows. You see no evidence of humans having cut these extensive benches along the valley walls or having formed them by fill from the surrounding terrain. What a massive project either would have been! And for what purpose? You observe that the stream presently occupies a wide, flat, valley floor—similar to the flat-topped benches—and the sediment in the stream is of sand and gravel. These benches are likely linked to the past behavior of the stream.

Reading about the geology of streams (see Chapter 2), you learn that the flat-topped benches in question are stream terraces—remnants of former valley floors or floodplains into which the stream has incised (see Figures 2-4, 9-15). Possible causes for incision are an increase in stream discharge—possibly from a change in the regional climate or overgrazing of the slopes and greater runoff from the watershed—or an increase in the gradient or slope of the stream bed. The gradient might be increased by the lowering of a feature controlling base level, such as a reservoir, natural lake, or resistant rock layer. Or the gradient might be increased by uplift of the region. Uplift? Hmm! Unable to relate to this, you decide that uplift need not be considered seriously.

Back to the valley, you observe the terraces more closely, and this time you take notes on what you observe. You've brought along a detailed topographic map of the valley and begin outlining the terraces on it. Armed now with several ideas for the origin of terraces, you apply them to "your" valley and begin to test them, one by one. Your favorite—an increase in gradient by a lowering of base level—is shot down when you cannot substantiate a former higher base level downstream. Other modes of genesis succumb similarly to your rigorous analysis until you are left with one: uplift, the seemingly implausible explanation.

What you have exercised for this geological problem is the scientific method: observing and gathering facts, and formulating and testing hypotheses to fit those facts. The scientific approach can be elaborated by the following steps (not necessarily in the order in which they are pursued):

1. State the problem clearly.
2. Reconnoiter the area or situation.
3. Review any previous study on the problem by others (this may precede the reconnaisance).
4. Observe, collect, and record facts.
5. Compile and synthesize the recorded facts.
6. Formulate multiple working hypotheses (this may happen spontaneously shortly after your becoming aware of the problem; it is best to always formulate more than a single hypothesis).
7. Test the hypotheses.
8. Select the hypothesis that best interprets the gathered facts.

There you have it. The geological approach to a geological problem is really *the* scientific approach.

But you've been wondering: What caused the uplift resulting in the formation of the stream terraces? Whoops! You're snagged by *another* geological problem! And the process of the geological approach begins once again.

20 HOW TO READ GEOLOGIC MAPS AND CROSS SECTIONS

Most maps locate places. But they may also locate products and plants and animals, and portray land surface configuration, as do topographic maps (see Chapter 1). But **geologic maps** are different; they portray rocks. Besides occurrence and extent of rock types, though, these maps indicate any deformation as well as age relationships. Geologic maps commonly exhibit the so-called *bedrock* as if loose, surface sediment and soil were stripped away. On others, the surface geology is realistically shown with all surface materials intact. Some surface geologic maps may evaluate environmental hazards; others, the engineering properties of rock materials. Where are landslides likely to occur? Where should you construct sewage lagoons? At what depth can you expect to find the water table? Other geologic maps concentrate on large-scale deformational features—folds, faults, joints—the *tectonic maps*. And *paleogeologic maps* portray land surfaces at former times.

Cross sections show us how rocks would appear at depth along the side of an imaginary trench cut straight down through the earth's surface; they are similar to the side view observed when a layered cake is cut. Cross sections complement what we can see on geologic maps.

Geologic maps and cross sections, most importantly, allow us to predict: to predict what we will find along newly built roads, excavations, tunnels, and in wells. And so they are particularly useful in locating oil, water, coal, iron ore, and other materials vital to people.

Reading Geologic Maps and Cross Sections

Let's take a look at a geologic map and cross section (Figure 20–1) to understand how they are read. Rock and sediment layers and bodies shown on geologic maps are those readily identifiable and traceable. If not traceable, they cannot be mapped. What is actually drawn are the boundaries or **contacts** between **formations,** the most frequently mapped rock or sediment units. Subdivisions or groups of formations are also occasionally mapped. Formations are emphasized on the map by color, pattern, or both, and labelled by a shorthand symbol. That which holds up V-shaped Cove Mountain (Figure 20–1) is the largely sandstone Pocono Formation; its symbol reflects its age as well as its name: *Mpo; M*=Mississippian, *po*=Pocono. Age relationships are summarized along a side or on the bottom of a map by a display stacking up all the rock or sediment units in their proper order with the oldest at the bottom. Formations are named, in North America, after places at or near where they were first identified. So the overlying and younger Mississippian Mauch Chunk Formation (*Mmc*), of shale, siltstone, and sandstone and forming a triangular lowland below and within the arms of Cove Mountain, was named after Mauch Chunk, Pennsylvania.

Rock deformation or geologic structure—or the lack of it—is clearly delineated on geologic maps. Turn back to Figure 20–1. The banding of the formations indicates tilted rock layers and reflects parallel ridges and intervening valleys (see Chapter 9). Several formations parallel V-shaped Cove Mountain and the Pocono Formation that holds it up. This configuration is part of the zigzag pattern of ridges and valleys (and formations) of plunging folds (see Figure 9–5). But, on the map, is this a plunging anticline or syncline? Of course, a dashed line, running along and near the middle of the Mauch Chunk Formation, says "Cove Syncline." But let's check this: The easiest way is by glancing at the cross section, which displays a distinct syncline. We can also identify the fold from the map. A V-shaped notch (see Figure 9–1) above point A causes indentations of formation contacts that point and dip toward the fold axis, only true for a syncline. Another clue to the type of fold relates to the relative age of the eroded formations. In eroded synclines, relatively younger formations are closer to a fold axis or center line as seen on a map; remember that Mauch Chunk—along the fold axis—is younger than Pocono. The opposite is true for anticlines. Notice how the stream drainage reflects the structure. This is part of a trellis drainage pattern (see Figure 9–9). Near the south (bottom right) edge of the map is an east-west line—heavier than a contact line—with *D* (down) on the north and *U* (up) on the south, reflecting relative movement along a fault. The cross section shows this as a high angle reverse fault (see Chapter 9). Extremely narrow, dark rock bodies transect the formations of the plunging syncline; these are basaltic dikes, depicted on the cross section by heavier vertical lines.

From geologic maps and cross sections, too, you can read the geologic history of a region. First, you assume that the folded beds (Figure 20–1) were originally flat-lying. From fossils we know that the oldest formations were laid down in the sea but those north of a line about halfway between Cove and Little Mountains were deposited on land. This change in formation origin implies uplift of the land, dropping of sea level, or perhaps some combination of the two. Some time after the Mauch Chunk was laid

FIGURE 20–1. Geologic map and cross section (along a line from A to the bottom of the map) of strata within a plunging syncline cut by basaltic dikes in south-central Pennsylvania. Beds are faulted near the southern edge of the map. (From *Pennsylvania Geological Survey Atlas* 137 cd by J. L. Dyson, 1967.)

down in streams and lakes, all the formations were squeezed together and folded, forming the plunging syncline. Since the basaltic dikes transect the Mauch Chunk, they must have been intruded after folding (see Figure 11–2). From the cross section we can tell that faulting followed folding, but did it occur after or before the intrusion of the dikes? We would require a fault in contact with a dike to determine this. After faulting and dike emplacement, stream erosion dissected the land surface, forming linear ridges in resistant sandstones and linear valleys in weaker shales and limestones. The Susquehanna River, though, cuts *through* resistant ridges like Cove Mountain, and its downcutting must have kept up with uplift of the land after folding. Besides downcutting, present streams have deposited sediment in their valleys.

Constructing Geologic Maps and Cross Sections

Let's say you've spent considerable time roaming central Pennsylvania and decide to try your hand at making a geologic map somewhere in that region. You select a relatively small area of about 6 square miles (16 square kilometers)—the smaller the better for a first attempt—straddling New Lancaster Valley (Figure 20–2). You obtain a topographic map (see Chapter 1) and use this as a base map; a good aerial photograph of a suitable scale could serve as well. First, you reconnoiter the region, find out how much walking will be required, where are the best exposures, and what rock types are present. Be prepared to expend some footwork. On the other hand, geologic maps can be constructed remotely from maps and photographs with strategic ground-checking (or very little or no ground-checking, as in the case of geologic maps of the moon or Mars). From the map and terrain itself, you realize the region is one of linear ridges and an intervening linear valley. These landforms suggest folded or fractured rock layers of varying resistance to erosion (see Chapter 9). Now, you get serious.

This being a region within a humid climate, vegetation obscures much of the rock; and the ridges, particularly, are heavily forested. Your best bet is to search first for exposures along gaps in the ridges, such as those at Reeds Gap and Kearns Gap, and another between New Lancaster Valley and Jack Mountain. You discover at Reeds and Kearns Gaps that Knob Ridge is held up by a sandstone with conglomerate. (You learn later that the same rock holds up a similar ridge across New Lancaster Valley.) At each good exposure you record characteristics of the rock: rock type, color of freshly broken and weathered surfaces, predominant minerals, any fossils and their abundance, and anything else that enables you to consistently identify it. Pay attention to the vegetation as it may reflect a particular rock type. Accurately plot the examined exposures on the field map with rock symbols and sketch in formation contacts where possible (Figure 20–3). (Normally you would plot directly on the topographic map. A

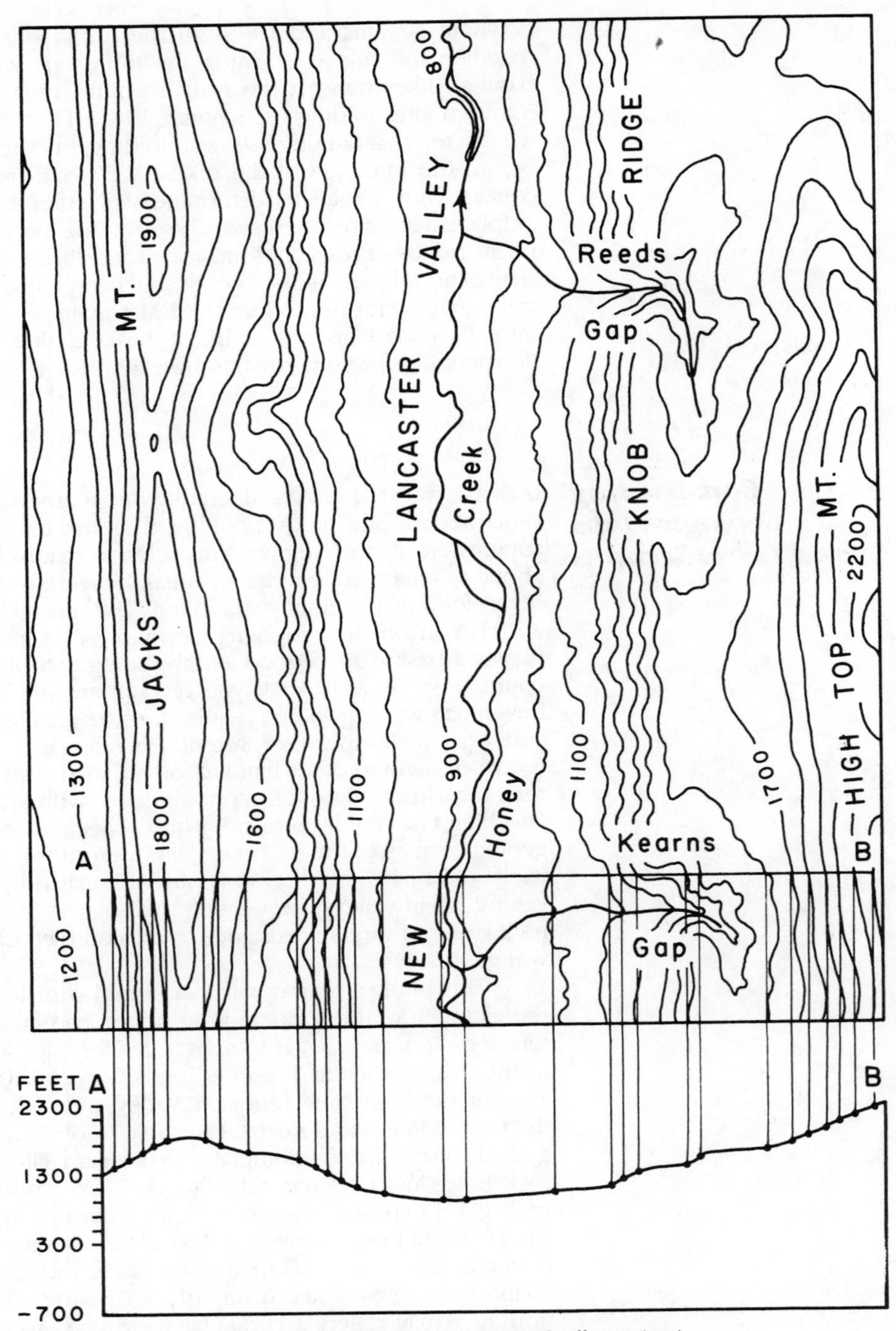

FIGURE 20–2. Topographic map and profile of a ridge and valley region in central Pennsylvania. For the profile along line *A–B,* drop a point on the elevation scale below wherever line *A–B* crosses a contour. Scale of the map is given in Figure 20–3. (Map is modified from *Pennsylvania Geological Survey Atlas* 126 by R. R. Conlin and D. M. Hoskins, 1962.)

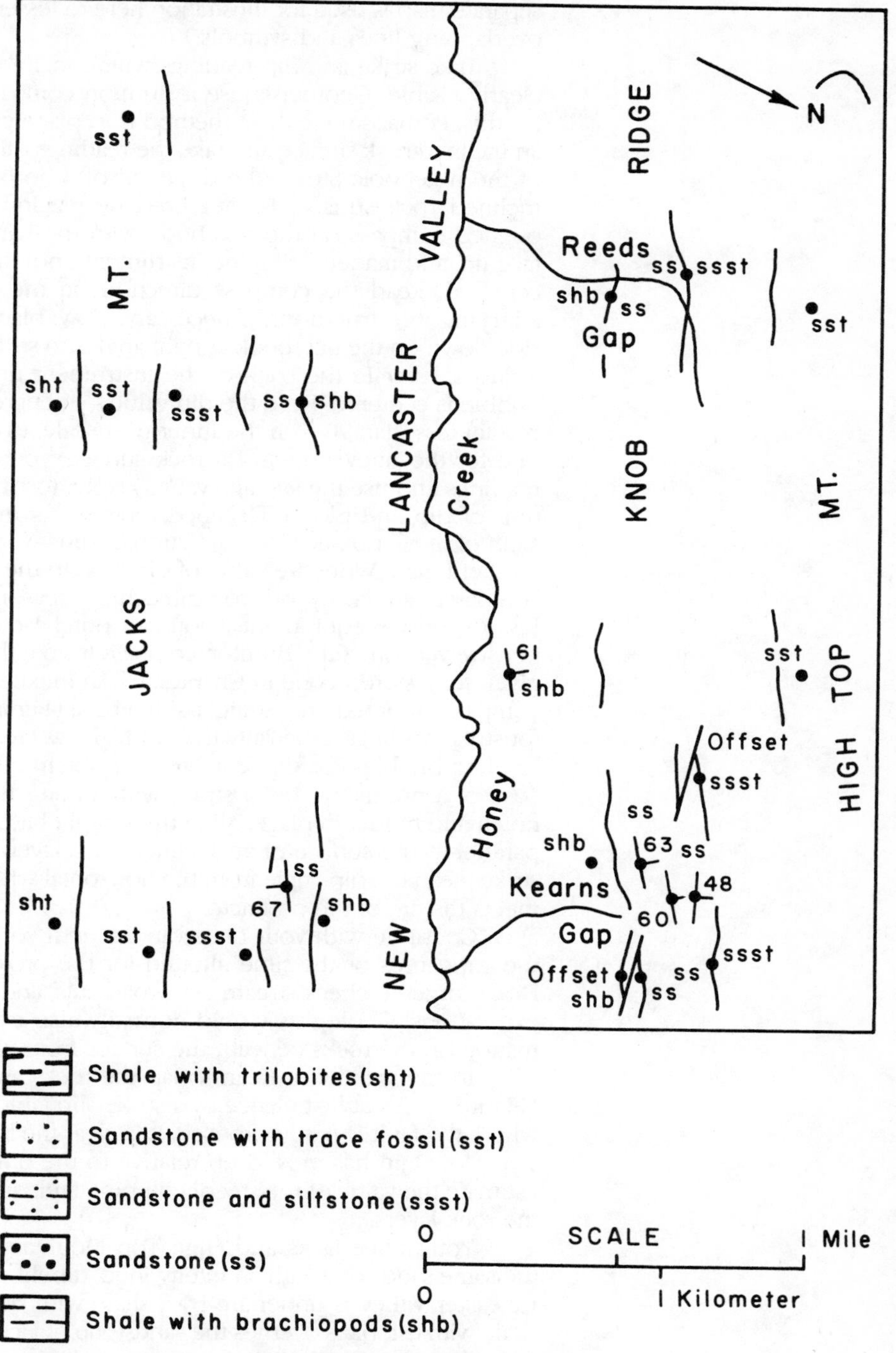

FIGURE 20–3. Hypothetical geological field sketch map of the region shown in Figure 20–2. (Based on data modified from the same source as for Figure 20–2.)

separate map is used for illustration here to lessen confusion from overlapping lines and symbols.)

Take strike and dip readings where inclined rock layers are clearly visible. Geologists use a Brunton compass (Figure 20–4) for this. Perhaps the easiest method is to place your notebook on an inclined rock surface and take the readings off the even surface of the notebook. Strike, the direction of a horizontal line on an inclined rock surface, is measured by placing an edge of the opened compass on the notebook with the window and mirror face up, and maneuvering the instrument until the level bubble is centered. Read the compass direction on the outermost scale, adjusting it to true north if necessary. Now, place the instrument side down on the notebook at right angles to strike (Figure 20–5). Adjust a lever in the back of the instrument until another level bubble is centered. Read the dip value—in degrees from the horizontal less than 90—on the innermost scale. (To partly compensate for the unevenness of a rock surface, you might take three readings and use the average value.) Note, too, the general direction of dip and plot a T-shaped dip-strike symbol—the shorter stem of the T points in the direction of dip—with a protractor on the field map. Write the value of dip next to the symbol. Brunton compasses are costly, but you can achieve reasonable results with less expensive equipment. (You can purchase a relatively inexpensive version of the Brunton compass from either of the sources given for a stereoscope in Chapter 1.) An inexpensive, fluid-filled compass mounted on a straight-sided base (Figure 21–1) is useful for strike readings, especially if an air leak has created an automatic leveling bubble. For dip, cut or file a notch at the center of the base of a protractor. Lay a string with an attached weight in the notch and hold it in place. Align the straight base of the protractor parallel to the surface of an inclined rock layer at right angles to strike. Read the dip angle from the horizontal where the taut string meets the arc of the protractor.

Continue with your observations until you've examined all the exposures or the time allotted for the project is consumed. Don't forget to check stream cuts, road cuts, and other man-made excavations. Geologists would normally also examine any information on the rocks beneath the surface from water or oil wells.

In the vicinity of Kearns Gap the rocks are distinctly offset (Figure 20–3), at first glance by a strike-slip fault. But at one place where the fault is partly exposed, it is clear the block toward High Top Mountain has moved up relative to the other. You can only estimate the value of the steeply dipping fault, greater than that of the rock layers.

You notice Jacks and High Top Mountains are held up by the same sandstone with an intertwined, tubelike trace fossil; New Lancaster Valley is underlain by a shale with brachiopods; and a shale with trilobites overlies the sandstone of Jack Mountain. Other areas intermediate in relief between prominent ridges and the valley are underlain by the same sandstone and siltstone formation.

FIGURE 20–4. One method of measuring strike of a sandstone bed, using a Brunton compass, by placing the instrument on a notebook and leveling it. The sandstone surface dips to the lower right at right angles to strike.

FIGURE 20–5. One method of measuring dip, using a Brunton compass, on the same sandstone surface as in Figure 20–4. The dip value in this case, read off the inner scale, is 14°. Note that the level bubble at the top of the instrument face is centered.

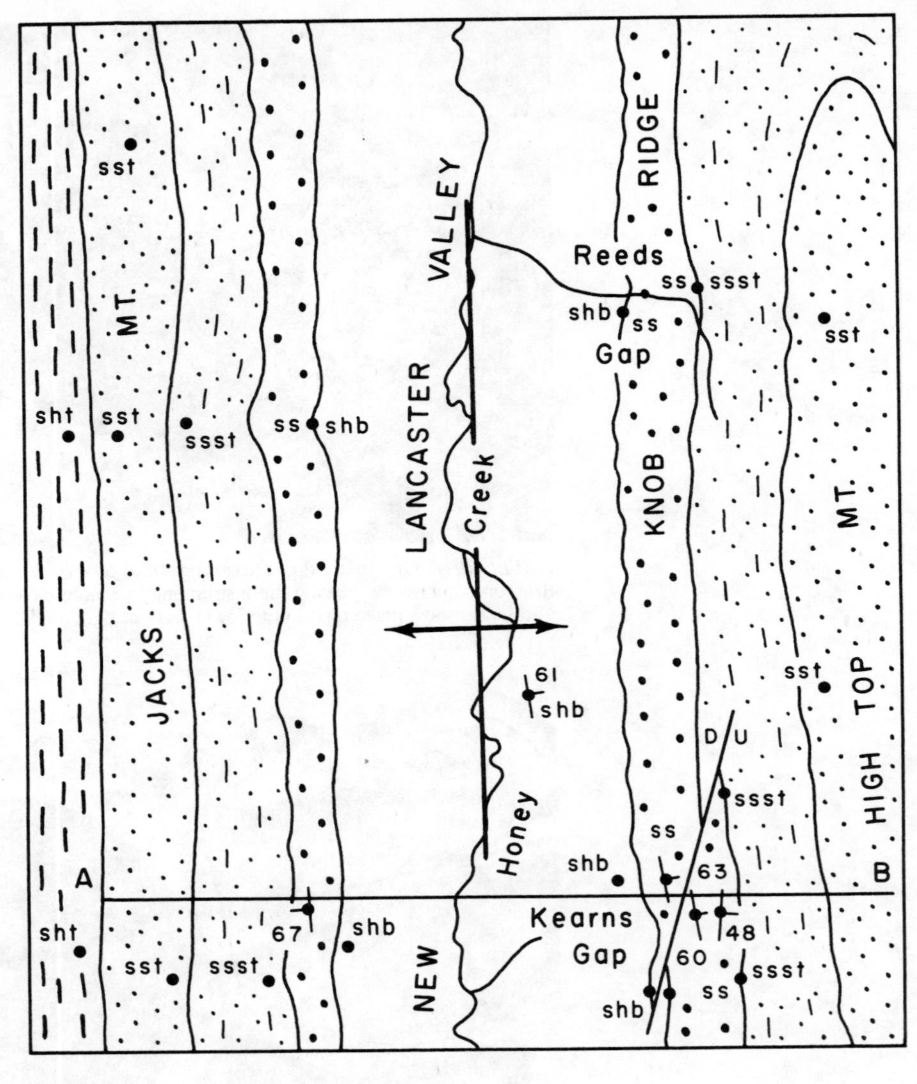

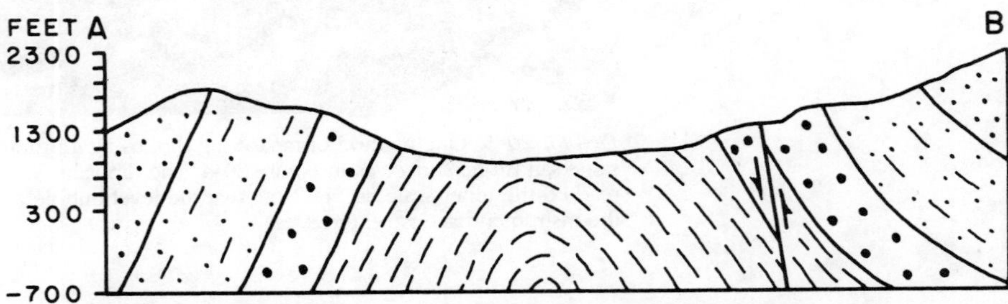

FIGURE 20–6. Completed geologic map and cross section based on Figures 20–2 and 20–3. For the cross section, drop a point on the cross profile wherever line *A–B* crosses a formation contact. Rock symbols are explained and arranged according to relative age in Figure 20–3. (Based on data modified from the same source as for Figure 20–2.)

Using the topographic contours as guides, you draw in all contacts. At ridge gaps they indent away from New Lancaster Valley, following broad V-shaped notches (see Chapter 9) that point in the direction of dip of rock layers. The fault is drawn with a heavier line than for contacts. Fill in the formation areas with rock symbols or formation patterns or colors.

It is clear you have five rock units that may be called formations (Figure 20–6). Recalling which formation underlies another, you arrange them all in their correct order, the oldest at the bottom (Figure 20–3). If time allows, you might try to identify the fossils and assign a real age to the formations.

Now, draw a cross section at right angles to the strike of the formations where you have the most dip-strike readings. A logical choice is along line *A–B.* First, draw a cross profile from the topographic map (Figure 20–2). Set up a vertical scale along one margin for elevations; it may or may not equal the horizontal scale. If not, the vertical dimension is drawn upward unnaturally or exaggerated. Wherever a contour line intersects line *A–B,* project its value straight down onto the vertical scale below and plot the position with a point. Connect all points with a smooth, curved line, keeping in mind the contour interval when extrapolating above the highest or below the lower points. Project now on the cross profile every formation contact that intersects line *A–B.* With a protractor, plot the dip values, watching for the correct dip direction. Maintain the formation thicknesses as you extend them deeper. Caution: Don't extend your cross section any deeper than the observations clearly warrant. From the opposing dips, it is clear that an anticline dominates the region, its axis or centerline—shown by a heavy long line crossed by a shorter double-barbed line—running along the center of New Lancaster Valley. High Top Mountain rests along the axis of an adjoining syncline. Show the formations in the cross section with rock symbols, aligning any linear symbols parallel to the layering. Depict relative movement on opposite sites of the fault with opposing arrows.

There, the project's done (Figure 20–6). And you have the feeling of knowing an area really intimately. Sometime later a geologist friend shows you a published geologic map that includes your mapped area. Your map is very similar to that shown on the published version—you did a good job! Your formations, from oldest to youngest, are called Reedsville, Bald Eagle, Juniata, Tuscarora, and Rose Hill; the first three are Ordovician in age, the other two Silurian.

By the way, it's not difficult to find out what areas have been mapped geologically and the sources of those maps. Simply write for a Geologic Map Index for the state in question from: National Cartographic Information Center, U. S. Geological Survey, 507 National Center, Reston, Virginia 22092.

21 HOW TO MAKE A MINERAL, ROCK, OR FOSSIL COLLECTION

Why Collect Minerals, Rocks, or Fossils?

Minerals, rocks, and fossils, I said earlier, are the stuff of geology—that of which the earth is made. A collection of these materials allows you to handle them frequently, pore over them, become intimate with the basic components of your natural physical surroundings. Collecting such materials takes you to the natural world outdoors and to places not regularly tramped.

Although professional mineralogists, petrologists, and paleontologists enjoy most discoveries in their field, an untrained collector can unearth significant finds as well. And amateurs have made important discoveries. They require mostly careful observation and perseverance. Then, too, collecting and maintaining a collection—a fun, interesting hobby in itself—may ultimately lead to a career in the natural sciences or related fields.

Collecting

Where to look

Whether you pursue minerals, rocks, or fossils, the general approach is similar: You seek out exposures of rocks, natural or man-made. Natural exposures include stream cutbanks or valley walls, cliffs, and caves or hillsides—particularly where vegetation is scarce. Where soil, sediment, or ice is lacking, rocks may be exposed in terrain of little slope. Man-made exposures are more numerous: quarries; pits; mines; excavations for dams, buildings, pipelines, and cables; tunnels; canals; and road and railroad cuts. Many such exposures can be located on detailed topographic maps. Quarries, mines, and gravel pits, especially, are plotted on these maps. Don't neglect geologic maps (see Chapter 20 and Appendix C) in your quest. For known specific localities, you can check collectors' magazines or publications of state geological surveys, professional geological societies, or the U. S. Geological Survey. Much of J. E. Ransom's *Fossils in America* (cited fully in Chapter 20) is devoted to collecting localities in the United States. More directly, ask local collectors, museum personnel, or geology professors. Most are willing to share collecting localities if they know you are a serious and conservative collector.

Collecting equipment

Collecting equipment (Figure 21–1) is relatively simple: a hammer, a chisel or two, a hand lens, newspaper, tape, knapsack or collecting sack or bag, notebook and pencil, slips of paper for labels, and perhaps an acid bottle. The hammer helps free a specimen from an exposure and enclosing rock. A geologist's or bricklayer's hammer is most often used. The pounding end of a geologist's hammer (Figure 21–2) is squarish; the chipping or picking end gradually tapers to a point—"hard rock" type—or chisel edge—"soft rock" type. A bricklayer's hammer (Figure 21–1) resembles the soft rock type, useful for splitting shaly rocks apart. The longer tapered end of either type of hammer facilitates prying out specimens from cracks and cavities. An extra, heavier, sledgehammer can be carried in the car for breaking up the occasional larger rocks. Cold chisels are handy for removing specimens when the hammer alone is insufficient. CAUTION: Watch out for flying rock chips when pounding, for yourself and others nearby. A good idea is to wear protective goggles, or, at least, turn your head slightly away.

A 5-to-10–power hand lens or magnifying glass is most useful for viewing mineral grains, small crystals, and small fossils or details of larger ones. Use the newspaper to wrap specimens—mainly to prevent them from rubbing and defacing each other during transport. Chip off unnecessary sharp edges from a specimen and lay it on a few thicknesses of newspaper, ripped to appropriate size. Insert a locality label before passing a corner of the newspaper over the specimen and snugging it under. Now fold the newspaper

FIGURE 21–1. Basic equipment for the amateur geologist on the run: clockwise, from upper right, are hammer, dilute hydrochloric acid, compass, hand lens, and protractor.

tightly against the specimen, left to right and right to left, folding further any small projecting corners. Roll the specimen tightly toward the remaining corner and fold this in toward yourself when you reach it. A small piece of tape—any will do, but filamentous is strongest—will seal the neat, tight package.

A pertinent rule for the serious collector might be: "If you haven't recorded the locality for a specimen, don't collect it." Similar is the adage: "A specimen is as good as its label." Comparison of specimens from different localities is meaningless without good locality data, and a collection donated to a university or museum will have limited use if lacking such information.

In your field notebook record the locality as precisely as you can: ideally so another collector—reading your description—could find the spot. Practically, this is not always possible. A relatively complete locality description might read: "East road cut South Dakota (Highway) 63, 16.4 air miles southwest of Timber Lake, South Dakota." (Relating road cut exposures to mileage markers results in an even more specific location.) Access to a map with legal land divisions allows you to prefix this description with "NE¼NE¼SW¼ Sec. 32, T. 15 N., R. 24E.", specifying the locality further. Other entries in the notebook include date, an estimate of abundance of the collected object—abundant, common, rare—specific occurrence in an exposure, and associated rocks, minerals, or fossils.

Assign a field number to each collecting locality for easy reference to it. Place this number in your field notebook and next to an accurate plot of the locality on a map. If rushed, the field number can substitute for a description on the locality label tucked in with a specimen. It is convenient, too, to write the number on the outside of a wrapped specimen. A simple system is a two-digit number for the year of collecting followed by a consecutive number for each locality during that year: 85-17. If necessary, the number may be modified for consecutive specimens from a single locality—85-17-1, 85-17-2—as where specimens are collected from different rock layers.

How to collect

Don't rush when collecting; be prepared to expend *time.* Examine an exposure systematically, left to right, bottom to top (if not too steep). But first, look over weathered rock accumulated at the base of a slope. Turn over rock fragments frequently. Look for associations of minerals and fossils with particular rock types. When you find a desired mineral, rock, or fossil, trace its source upslope.

Minerals and rocks are best sought from fresh exposures, little altered by weathering processes. You don't want iron oxide-stained or partly dissolved specimens for your collection. Detect translucent or transparent minerals in rock rubble more readily by walking into the sun. Agates, for example, are easily seen along beaches or in dry stream beds this way.

Fossils can be satisfyingly taken from both fresh and weathered exposures. Those preserved as carbonized films or thin

FIGURE 21–2. Cretaceous ammonoid cephalopod broken out from a concretion with a geologist's hammer.

impressions, such as leaves, are best collected by splitting fresh shales. Others, such as brachiopods or blastoids with relatively thick hard parts, readily free from shales and limestones by weathering and thereby are easily collected. Fossils are generally best preserved in shales and limestones; shales, in particular, allow for the most detailed preservation. Break apart concretions for fossils (Figure 21–2). For serious study collect also the impressions in the enclosing fragments. Commonly, concretions are jammed with fossils, whereas the enclosing rock may be barren or rarely contain them. If you come upon part of a vertebrate skeleton that seems to continue into the exposure (Figure 21–3) *do not* begin digging immediately! Think through the situation carefully. If you are untrained in vertebrate fossils and their excavation, a good approach is to notify a vertebrate paleontologist at the nearest university or museum. You might submit a few bones—noting their position—that are well below but presumably derived from the in-place skeleton. If possible, submit a few photographs as well. You might have discovered a significant find that a trained and well-equipped expedition can completely and properly excavate for later study and display.

As you work an exposure, set specimens aside; always attempt to improve the lot. When finished at an exposure, select only the best specimens from those accumulated to become part of your collection.

Crumbling or flaking fossils—or the matrix they are in—may have to be impregnated or hardened before collecting. Readily available Elmer's glue—easily thinned with water—may be used

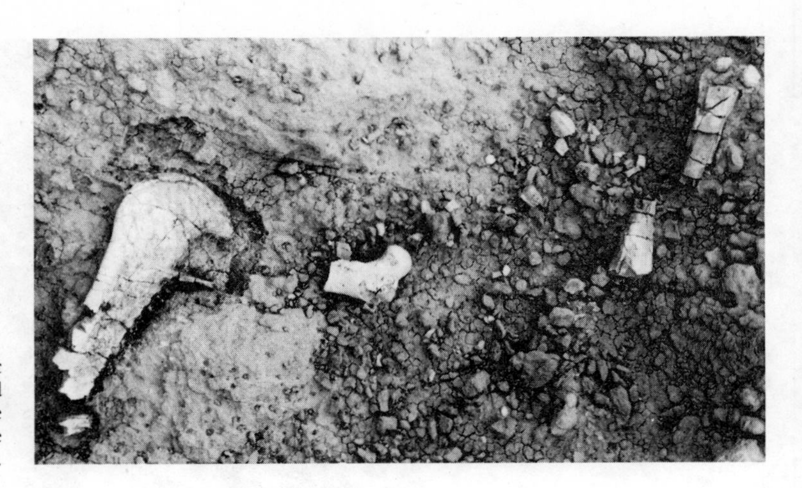

FIGURE 21-3. Middle Cenozoic mammal bones partly weathered out of mudstone. Length of the incomplete lower jaw at the left is about 6.3 inches (16 cm).

for this purpose. Duco cement—soluble in acetone—is another possibility. Spraying with a plastic spray may be just the thing for flaking fossils.

Fragile specimens may be wrapped in toilet paper and packed in a rigid box or can. They too, may be left in matrix and removed slowly and carefully at home.

Ethics of collecting

Collecting ethics include at least the following:

1. Always make a reasonable attempt to gain authorized access to any property, private or public. If you lack permission to enter a property, you are violating the common law of trespass. Landowners may show reluctance to grant you permission because of their responsibility for any injury you might sustain, unless their state has an owner-release law relieving them of such responsibility. Collecting on federal land, except in national parks and similar prohibited areas, is generally allowed for the noncommercial, avocational collector taking a few specimens without the use of motorized excavating devices. Collecting on most state land is similarly little restricted for the avocational collector. A few states, however, have specific regulations, and an appropriate state agency, such as a state geological survey, should be contacted for information on such regulations.

2. Once your entrance has been authorized, behave on property as the honored guest you are. Close any gates you have opened. Do not disturb crops, buildings, or other personal property. Remove litter—even some you didn't create. Use every precaution against starting fires. If you must excavate, fill in all holes. And, particularly on private property, avoid the use of alcohol and drugs and minimize unnecessary noise. In the hope of future visits by you and others, as well as in fairness, you might return the landowner's favor with a gesture of your own: present him or her

with an unusual specimen from your collection, a book, food or drink.

3. Use restraint in collecting. Specimen-hogs are as bad as fish- and game-hogs. Collect only what will enhance your collection and what you can truly use. If a specimen will decorate the mantelpiece for a time, but will eventually be thrown out, leave it at the exposure. And don't collect specimens for later sale. This leads naturally to overcollecting and exploitation—profiting from a landowner's favor.

Preparing Specimens

Unless the specimen is fragile or crumbles in water, a good washing—with detergent—is usually necessary. Scrub it with an old toothbrush. Stubborn dirt may require considerable soaking.

Additional unwanted rock matrix can be removed with a hammer and chisel. Thinner or slabby parts may be nibbled away with pliers, or sometimes cut off with an old hacksaw. Small amounts may be removed carefully with resharpened and modified dental tools and crochet hooks. Scrape and flake slowly and carefully, and frequently wash the cleaning area.

Unwanted carbonate minerals or rocks may be dissolved by acetic or hydrochloric acid, variably diluted with five to ten parts water. Always test a small fragment first, however, before deciding on the type and strength of acid. Acetic acid, the gentler of the two, is present in vinegar; concentrated glacial acetic acid—used in the stopbath for photograph processing—can be picked up at a photographic supply store. Hydrochloric acid is known commercially as muriatic acid and is used in the geologists' acid bottle. Acid dissolving—to be done always in a nonmetallic container—works best in cleaning and freeing siliceous materials. More detailed information on the preparation and cleaning of fossils is given in *Fossils for Amateurs, A Guide to Collecting and Preparing Invertebrate Fossils* by Russell P. MacFall and Jay Wollin.

Identifying, Cataloging, and Displaying Specimens

Chapters 14 to 18 help you start identifying minerals, rocks, and fossils. If the collection bug bites hard, you will want to consult your nearest library for the numerous publications available or request interlibrary loans from a university or college library. Joining a mineral, rock, or fossil club will help with some of your identification problems as you interact with those having larger collections. And university, museum, and geological survey staff are usually willing to help if you request only occasional aid and don't overpower them with specimens.

Cataloging is the assigning of numbers to collected and prepared specimens; the numbers are tied to a running list. If a label is lost or the specimen is displayed without one, such a specimen can be readily identified and its locality recovered. The easiest and most readily used system assigns "1" to the first

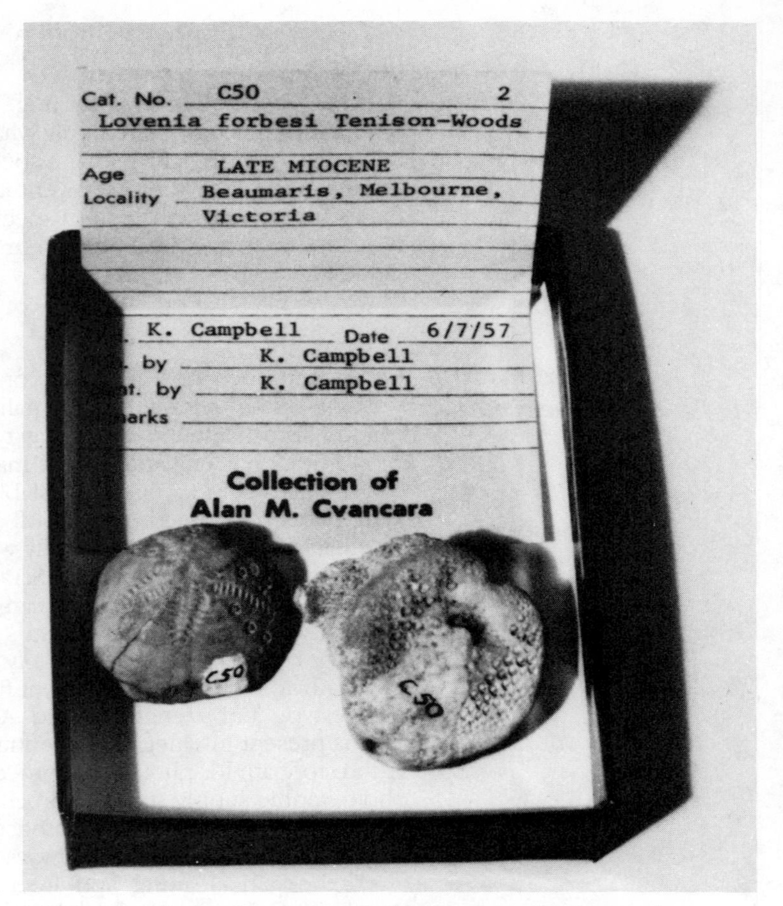

FIGURE 21–4. Cataloged specimens of a Miocene sea urchin in a specimen tray with a label. The catalog number (*C50*) may be placed directly on a specimen (right) or over a dab of white paint (left).

specimen and numbers others consecutively. Actually, one or more specimens from the same locality can be assigned the same number to minimize numbers (Figure 21–4). The catalog can be a book or a card file, which works better for larger collections. Such a file may actually consist of two: one alphabetic, the other numerical. If you replace specimens with those of better quality, replacing cards is relatively simple. The catalog bears what is on the specimen label but perhaps in more detail: catalog number and number of specimens, name, perhaps age and rock formation, locality, collector, date, and perhaps donator, identifier, and other remarks.

Print the catalog number in an inconspicuous place on the specimen in waterproof India ink. On rough surfaces dab on a spot of white paint first.

Visit the homes of fellow club members or museums for ideas on displaying specimens. Glass-fronted display cases or nonglass cabinets with shallow drawers house most specimens. In display cases, specimens frequently rest on styrofoam, plastic, or wooden bases, and may be variously lit. In cabinet drawers, specimens are placed in trays with their labels (Figure 21–4). If the label is folded at 90°, with the catalog number and name on the vertical part, it is easy to locate a specimen from the front of the drawer.

Organizing specimens within a collection depends largely on personal preference. But minerals are frequently arranged by their chemical composition, such as sulfides, oxides, and silicates; rocks by their classification, such as conglomerates, sandstones, and shales; and fossils by group, such as sponges, corals, and bryozoans. As you specialize with your collection—most collectors do—your organization will become more specialized as well.

Selected Readings

MACDONALD, J. R. *The Fossil Collector's Handbook: A Paleontology Field Guide.* Englewood Cliffs, N. J.: Prentice-Hall, Inc., 1983.

MACFALL, R. P., and JAY WOLLIN. *Fossils for Amateurs, a Guide to Collecting and Preparing Invertebrate Fossils.* New York: Van Nostrand Reinhold Company, 1972.

22 HOW TO SLEUTH STONE

A stone is not a stone is not a stone. To a geologist a **stone** commonly means a loose mineral or rock fragment shaped by natural processes; sand grains may or may not be included. R. V. Dietrich's interesting little book, *Stones,* is devoted to them; tidbits of information include examples of stones' therapeutic value, such as the rubbing of smooth stones for their tranquilizing effect. To one who sells stone, it is a natural hard rock material quarried or mined largely for constructional or industrial use and altered only by shaping or sizing. Gemstone, precious or semiprecious stone used as a gem when cut and polished, is part of this category. *Construction stone,* used directly for construction without chemical or heat treatment, includes crushed or broken stone and dimension stone. *Crushed* or *broken stone* is used for riprap on breakwaters, piers, and the like; concrete aggregate; cement; highway base; railroad ballast; and a myriad of other uses. *Dimension stone* is cut into blocks and slabs for a variety of purposes. Stone. Rock. Rock to the stone industry is stone still in place. To a geologist it is an aggregate of minerals (see Chapter 14) or a relatively hard, coherent, naturally formed mass of minerals.

I hope this stone-rock discourse has enlightened and not confused you, for such confusion is frequent. In the Viking cartoon, *Hagar the Horrible,* a jewelry merchant attempts a sale to a reluctant Hagar for his wife. Persistently, the merchant asks Hagar for his wife's birthstone. Hagar, presumably conceiving a hard, strong, appropriate "stone," replies: "Granite."

Stone sleuthing, the tracking down of the source of stone utilized by humans, is a conscious or subconscious pastime of many geologists. It may never become an overwhelming craze, but it intrigues those who favor a little detective work now and then. And you always learn a little geology rather painlessly in the process. Let's restrict our sleuthing to dimension stone because the pieces tend to be larger, and cut or polished surfaces are more readily observable. But first, let's examine the kinds and uses of dimension stone.

Kinds of Dimension Stone

The rock classification of the dimension stone industry is very much simplified compared to that of a petrologist. Seven groups basically take care of it: granite, traprock or basalt, limestone, marble, sandstone, slate, and greenstone.

Granite includes all granitic rocks (see Chapter 15) as well as gneiss. Stretching the definition further—to a geologist—allows for the inclusion of diorite, gabbro, and similar dark rocks—the *black granite* of the stone industry. Color is of utmost importance in further naming, as in "Georgia Gray," "Mahogany," and "Wausau Red." *Traprock* or *basalt* encompasses all dark igneous rocks too fine-grained for black granite, but finer-grained diorite, gabbro, and similar rocks may, at times, be included. Even darker felsite, as andesite, may be called traprock.

Limestone generally agrees with the geologist's concept (see Chapter 16), but includes also dolostone. *Marble,* though, means commercially any limy, crystalline rock or serpentine that will take a polish. *Onyx marble* or *Mexican onyx*—most comes from Mexico—is dense, banded, translucent, and unmetamorphosed and formed from cold-water precipitation. If dripstone (see Chapter 6), it's called *cave onyx. Travertine,* to geologists a variety of cellular limestone, is called marble if banded and dense enough to accept a polish. Green to nearly black *verde antique* or *serpentine marble* consists of serpentine (see Chapter 14) crossed by small veins of lighter minerals. It's called marble because of the marble-like veining and capability of taking a good polish.

Sandstone commercially also encompasses the conglomerate, breccia, and siltstone of the geologist; *quartzite,* too, may be included with sandstone. Other names used by the stone industry are: *puddingstone,* colloquial for certain conglomerates; *bluestone,* a hard, commonly dark gray, feldspar sandstone that splits easily into slabs; *brownstone,* a brown to reddish brown feldspar sandstone, typified by those from the northeastern United States; *freestone,* one that slabs in any direction; and *flagstone,* a sandstone or slate that splits into large, thin slabs.

Slate, a fine-grained rock readily cleavable into thin sheets, is essentially as the geologist conceives it. *Greenstone,* also an old field term used by some geologists, is a dark, crystalline metamorphic rock that derives commonly from the metamorphism of basalt, and is green because of such minerals as chlorite and epidote.

Many other varieties of stone may be used occasionally for dimension purposes, the so-called *miscellaneous stone.* Examples include soapstone (rich in talc), pumice, fieldstone (of cobbles and boulders), pegmatite, and petrified wood.

Uses of Dimension Stone

The use to which a dimension stone is put depends on such qualities as strength, durability, hardness, and color. Strength relates to a stone's resistance to crushing or bending. A stone's durability is its maintenance of strength with time, and its resistance to weathering (see Chapter 23), impact, and abrasion. Hardness relates somewhat to durability but also to ease of workability or shaping. Dimension stone may be classified to use as building stone, ornamental stone, monumental stone, paving blocks, curbing, flagging, roofing slate, and millstock slate.

Building stone incorporates a large variety of stone in exterior and interior construction. This includes stone for bridges, sea walls, and the facing of levees and dams. Cost allows only the rare use of massive blocks in buildings today, but stone blocks and slabs in such places as cornices and trim around doors and windows are common. Thin veneer slabs may cover walls or parts of them.

Ornamental stone is valued for its markings or color. Most is used for interior finishing; marble in thin slabs is a favorite, but limestone, sandstone, granite, and slate are also used.

Monumental stone must meet more exacting requirements than building stone, and to some extent, ornamental stone. It should, for example, be uniform in color and texture, free of flaws, and allow a good contrast to lettering when polished. Another important quality might be the lack of such minerals as pyrite that may cause rust stains upon weathering. Tombstones fit into this category of dimension stone. A variety of monumental stone is *statuary stone,* with the added requirements of easy workability and not chipping when carved. Granite and marble constitute most of the monumental stone. White, fine-grained, uniform marble is the preferred statuary stone. A fuzzy line may exist between building and monumental stone. The Lincoln Memorial in Washington, D. C., for example, is a building of monumental stone.

Paving blocks are rectangular blocks originally used for areas subjected to heavy, abrasive traffic such as that in city streets and freight yards. Its use has largely diminished to patios, borders for walks and gardens, and the like. Granite is most often used.

Curbing is long, thin slabs for edging streets and roadways. Granite, quartzite, and certain types of sandstone provide the most resistance to abrasion and weathering, and so are used mainly for curbing.

Flagging or flagstone is of thin, easily split slabs for lining walks, patios, courtyards, and the like. Mostly sandstone and slate is used, but certain types of limestone may be suitable.

Roofing slate must be easily cleavable, smooth, and lack imperfections. Its use has drastically diminished because of the availability of other, cheaper roofing materials. But some still favor its attractiveness and unusual durability.

Millstock slate includes blocks and slabs for sills, switchboards, billiard tables, blackboards, and the like. It must be even-grained but not all that fissile (splitting readily into thin sheets). The best blackboard slate comes from Lehigh and Northhampton Counties, Pennsylvania.

Dimension stone is used for a myriad of other uses, but on a minor scale as compared to those already covered. A sampling: grindstones (sandstone); tubs, sinks, and tanks (soapstone, slate); shower and toilet stalls (limestone, marble); and fireplaces,

FIGURE 22–1. Church walls constructed (1949) of split fieldstone of glacial cobbles and boulders. Salem Lutheran Church, Hitterdal, northwestern Minnesota.

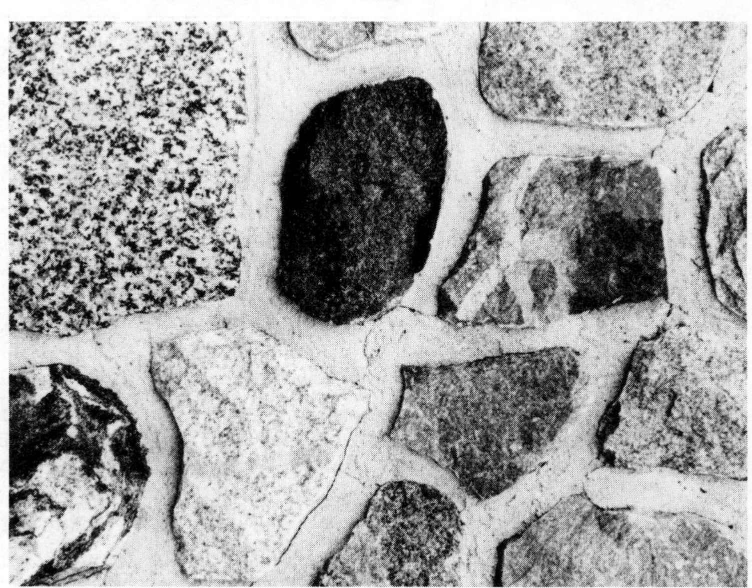

FIGURE 22–2. Close-up of fieldstone construction in the housing for notices of services near the church shown in Figure 22–1. Various metamorphic and igneous rocks are visible.

chimneys, and decorative walls—fieldstone (Figures 22-1, 22-2), pegmatite, petrified wood, slate and related rocks.

Practicing the Art of Stone Sleuthing

Armed with the knowledge of the main kinds and uses of dimension stone restricts the field somewhat. And, for the United States, knowing the main producing states and exporting countries (Table 22-1) helps narrow down possible choices in a general way. Yes, we have to keep in mind exporting countries, too—wherever we are. More than two and one-half decades ago, I stood at the edge of a small rural cemetery in eastern New South Wales, Australia. Pointed out to me was a dark, pillar-like grave marker rising above an overgrowth of bracken. Imagine my surprise when I learned the stone was laurvikite (or larvikite)—an essentially quartz-free granitic rock with pyroxene—from southern Norway! In the United States, granitic and metamorphic rocks are extracted from mountainous belts and nearby regions of the Appalachians, Rockies, and Pacific mountain belt. Granite of the Lake Superior region, chiefly Minnesota and Wisconsin, derives from eroded mountain roots in a southern extension of the Canadian Shield. Limestone and sandstone come mostly from interior states, where igneous intrusion and metamorphism are lacking, and interiorward parts of coastal states. Traprock is produced from a few eastern and western states having undergone, geologically, relatively young

Table 22-1. MAIN PRODUCING STATES AND COUNTRIES EXPORTING TO THE UNITED STATES DIMENSION STONE

DIMENSION STONE	*MAIN PRODUCING STATES AND COUNTRIES EXPORTING TO THE U. S.*
1. Granite	California, Georgia, Maine, Massachusetts, Minnesota, New Hampshire, North Carolina, South Dakota, Texas, Vermont, Wisconsin; Brazil, Canada, Finland, Italy, Japan, Norway, Sweden, United Kingdom
2. Traprock	Hawaii, New Mexico, Oregon, Pennsylvania, Virginia, Washington
3. Limestone	Alabama, Illinois, Indiana, Iowa, Kansas, Minnesota, Missouri, Ohio, Pennsylvania, Texas, Utah, Wisconsin; Canada, Italy
4. Marble	Alabama, Arkansas, Georgia, Missouri, North Carolina, Tennessee, Texas, Vermont; France, Greece, Italy, Mexico, Portugal
5. Sandstone (and Quartzite)	Arizona, Arkansas, Colorado, Indiana, Kentucky, Missouri, New York, Ohio, Pennsylvania, Tennessee, Utah, Wisconsin
6. Slate	Maine, New York, Maryland, Pennsylvania, Vermont, Virginia; France, Italy, Portugal
7. Greenstone	Virginia

extrusive igneous activity. Table 22-1 tends to reflect these generalities.

Now, for the actual sleuthing. First, identify the suspect to the best of your ability, consulting Chapters 14-17. Pay attention to accessory, or unusual, features such as bedding, mottling, foliation, fossils, and weathering effects. Take some time with this, and don't worry unduly if, say, passersby stare suspiciously as you snuggle up to the wall of a bank building with your hand lens. Don't trust your memory; take good notes. Photograph the stone if you have the equipment, watching for the best light—if you have the time—to emphasize the stone's qualities.

Once the suspect is identified, you proceed in much the same manner as a detective tracking down a human. Ask questions. Interview. A tombstone? Consult cemetery caretakers and dealers of tombstones. Most will probably be glad to help once they realize you are serious. Stone in a building? Interrogate companies that sell construction materials and construction firms themselves. Should this lead nowhere, locate the builder—perhaps through the owner if the two are not the same. For older buildings, where builders have left or are deceased and present owners lack knowledge of building materials, your job becomes more involved. You might then have to pore over city or county records. Don't become discouraged if the desired results take some time surfacing.

For a traveler, stone sleuthing is also possible, but answers may not come as frequently because of the less intensive search. Keep an eye out for dimension stone quarries, and attempt matches in those cases where a stone is distinctive even from some distance. Conversely, if you suspect a local source for a stone in a building, tombstone, or monument, ask residents about quarries and check them out. An example: You're traveling through north-central Kansas, sort of daydreaming, the car seemingly driving itself. After a time, something draws your attention to the mile-after-mile parade of fenceposts along the road. There's something different about them. Well, of course, they're of *stone!* Screeching to a stop, you run out and take a look. Tan limestone, with fossils—clams, ammonoid cephalopods (see Chapter 18). At the next town you discover the fenceposts came from local quarries in a bed of "fencepost" limestone used also for clothesline poles, flagpoles, tombstones, watering troughs, buildings, and other uses—primarily in the 1800s but later as well.

And now for a final, but significant point: Always check out any testimony or story about some source you have garnered. Don't antagonize, but analyze and confirm. A careful sleuth always does this.

Selected Readings

BARTON, W. R. "Dimension Stone," *U. S. Bureau of Mines Information Circular* 8391 (1968), 1-147.

DIETRICH, R. V. *Stones: Their Collection, Identification, and Uses.* San Francisco: W. H. Freeman and Company Publishers, 1980.

23 HOW TO READ ROCK WEATHERING FROM TOMBSTONES

You're bent over a tombstone and attempt to ignore the police car that approaches and stops within your peripheral range of view. The officer steps out, slams the door, walks over to you. A certain tense, empty feeling comes over you. What have I done now? Aren't cemeteries public places?

"Good afternoon," says the officer politely, but with an unmistakable look of suspicion. You nod, step back slightly from the tombstone.

"The caretaker says you've been spending quite a bit of time over here." So that's it, the caretaker reported me. "Uh, you wouldn't be one of those folks who get their kicks hanging around graveyards? Or maybe you're passing drugs out here? I'll have to see some identification."

You stare at the officer, dumbfounded. Hesitatingly you reach for your wallet and hand it over.

"Would you remove the cards, *please?* Hmm, this seems to be in order."

"Yes, officer, I am *totally in order.*" Your voice assumes a stern tone. "And I am *not* a ghoul or drug pusher! I'm simply checking these tombstones to see how they weather. You know, something like weather-beaten faces. And since tombstones are dated, you can see what changes have taken place over a specific interval of time. You see . . ."

"Uh, oh sure, I suppose so." The officer notices your hand lens and notebook for the first time. The look of suspicion dissipates. Clearly you are harmless. "Ah, well, best be on my way. Have a good day!"

Now, with that annoying encounter over, let's pursue rock weathering in general before examining tombstones further.

Rock Weathering and Its Products

Weathering is natural decay. We see its effects in paint cracking and peeling off buildings, bricks and concrete crumbling, and metals rusting and corroding. **Rock weathering** is the in-place breakup of rocks as they interact with the atmosphere, particularly with water, oxygen, and carbon dioxide. Given sufficient time, all rocks will succumb and crumble to form that most vital of earth materials that support life directly—namely, **soil.** Mixed with organic matter, soil differs from sediment that has been moved and later deposited or precipitated. Rock weathering provides the loose material from which sediments (and sedimentary rocks) are made.

Rocks weather physically and chemically. **Physical weathering** is the breakup of rock into smaller fragments by physical, nonchemical means. Chief processes are **ice wedging**—already mentioned in Chapter 3—and sheeting. Water seeps into cracks, expands its volume by 9 percent upon freezing, and wedges apart rock particles or blocks. In loose sediment, water freezing causes ice heaving, evidenced by tilted sidewalks and raised areas—later, holes—in roads. **Sheeting** or unloading breaks rocks into sheets or slabs, not by ice wedging but by release of pressure. Massive igneous rocks form at depth under high temperature and high pressure—confining pressure from all directions. Erosion removes the overlying rocks and lessens pressure. As the massive igneous rock is brought toward the surface, it expands upward by the release of pressure and forms fractures parallel to the surface. Quarry workers witness these fractures forming, often accompanied by loud noises, after rock is removed. Plant roots can wedge apart rock and sediment, and forest and grass fires may cause rocks to fracture and flake as they cool.

Chemical weathering consists of rock decay by changes in chemical composition and the internal structure of minerals making up the rocks. Rocks fragmented by physical weathering are set up for attack by chemical weathering: The smaller the particle, the greater is the surface area for chemical reaction. First, rocks weather chemically by dissolving, most notably the carbonate rocks limestone, dolostone, and marble. (This process was described in Chapter 6.) Rocks dissolved form rough surfaces, often pitted, grooved, or fluted (Figure 23–1). Since the beginning of industrialization and the expansion of fuel-driven vehicles, the burning of fuels has released tremendous amounts of carbon dioxide and sulfur dioxide into the atmosphere. These gases combine with precipitation to form carbonic and sulfuric acid, released as acid rain, fog, and snow—prevalent in such places as the northeastern United States. You can imagine how these added acids speed up the solution of rocks, as well as detrimentally affect the life of natural waters.

Second, rocks weather chemically by minerals combining with oxygen—oxidizing or forming oxides, such as hematite or limonite (see Chapter 14). Minerals rich in iron, particularly, oxidize. Rusting of iron is oxidation.

And, third, rocks weather chemically when minerals combine chemically with water or hydrate. Let's imagine, now, a granite hydrating as well as oxidizing to some degree. Remember granite

FIGURE 23-1. Fluted and pitted surface of dissolved limestone. The flutes at least partly follow fractures. The widest flute is about 0.6 inch (1.5 cm).

(Chapter 15)? A coarse, igneous rock mostly of feldspar plus quartz and ferromagnesian minerals—containing especially iron and magnesium, like biotite and amphibole. The feldspar combines with water and carbonic acid to form clay minerals, salts in solution (mostly carbonates and bicarbonates of sodium, calcium, and potassium) that are flushed away, and silica, some of which is removed in solution and some of which may remain behind as quartz. Likewise, the ferromagnesian minerals combine with water and carbonic acid to form essentially the same products as the weathered feldspar. The removed salts, however, are carbonates and bicarbonates of iron, magnesium, and calcium. In addition, oxidizing of iron produces iron oxides. The quartz grains of the granite, very resistant to weathering, may be rounded by slight dissolving, but remain behind as little-changed grains of sand. Think of it, a mighty rock—granite—reduced ultimately to a rusty, sandy clay soil! All it takes is air, water, and time. And there is plenty of all three.

Prior to soil formation, an intermediate stage of weathering in particularly coarse igneous rocks is **spheroidal weathering,** the production of rounded or spherical surfaces on weathered rock. Rounded or near-spherical boulders may even result. Such weathering is initiated by intersecting joints or fractures. Chemical weathering attacks the rock, particularly along fracture surfaces, and more so on the corners and edges of rock blocks. In time, the blocks assume a rounded or spherical form. A related process is **exfoliation,** the breaking of rock into concentric sheets, shells, or slabs. It may result from sheeting, ice wedging, or chemical weathering, or a combination of all three processes. In chemical weathering, the expansion caused by a volume increase as feldspar weathers to clay minerals may be of considerable significance. Exfoliation forms rounded hills called **exfoliation domes** (see Chapter 10, Hills), particularly well displayed in Yosemite National Park, California (see Figure 3–6). Intersecting joints in sedimentary strata also allow physical and chemical weathering to sculpture fascinating rock fins, arches, and columns or pinnacles. These features are well displayed in such places as Arches (see Figure 13–1) and Bryce Canyon National Parks, Utah.

What affects the rate at which rocks weather? Primarily climate, followed by rock type. Water is a vital ingredient in any kind of weathering, and higher temperatures speed up chemical reactions. So chemical weathering is fastest in wet, hot climates forming thicker soils and slowest in dry, cold climates where thinner soils result. In dry or cold climates chemical weathering is minor or insignificant, and physical weathering dominates. Temperate climates display intermediate rates of weathering and soil thicknesses, and a more even mix of chemical and physical weathering.

Rocks of rather soluble minerals, such as limestone, dolostone, marble, rock salt, and rock gypsum, weather—dissolve—relatively rapidly in wet climates, slowly or insignificantly in dry climates. So a limestone in New York will weather faster than one in Utah. Rocks of silicate minerals (see Chapter 14)—quartz, feldspar, olivine, and the like—weather in relation to how distant their conditions of formation were from those at the earth's surface. Dark minerals, such as olivine (least resistant to weathering), pyroxene, and amphibole, have crystallized from molten rock material under high temperatures and pressures. Light minerals, such as muscovite and quartz, have crystallized at relatively low temperatures and pressures. So, dark igneous rocks, now at the surface and far from their conditions of formation, weather much more rapidly than light-colored ones. Quartz is one of the most resistant minerals to weathering, and rocks largely of this mineral—quartzite, quartz sandstone—are, therefore, resistant too; unless, of course, the grains are cemented together by calcite or some similar soluble mineral.

Rock Weathering from Tombstones

What we can read from tombstones—and other similar kinds of dimension stone (see Chapter 22) used for monumental, ornamental, and building purposes—are *dated* effects and rates of weathering. As already mentioned, both effects and rates relate largely to rock type and climate.

Most tombstones are of granite and related rocks, marble,

limestone, sandstone, and slate; marble and granite seem to prevail today. (You may wish to check back with Chapters 15 to 17 if you've forgotten how to identify these rocks.) Marble, of course, weathers chiefly by dissolving, and more so in wetter climates. Solution is evidenced first by a dulling of polished surfaces; later, these surfaces become roughened, in extreme cases pitted and grooved. Roughening provides more exposed surface area and greater attack by further weathering. If lichens—incompatible with polluted atmospheres—begin growth on a marble tombstone, solution is enhanced. Lichens retain moisture in their vicinity and secrete acids. Tombstones receiving precipitation dripping from overhanging trees are also more susceptible to solution. And areas of frequent fog, particularly, dissolve marble, for water droplets remain suspended for a long time.

Try measuring the amount of surface removed by solution to arrive at a rate of weathering. Required for accurate results is a resistant, essentially insoluble mineral within the marble. A quartz vein, cut and polished within a flat surface of the marble and now standing in relief above a dissolved surface, is ideal. Professor E. M. Winkler measured, in the 1960s, such a vein at the top of a Vermont marble tombstone in humid South Bend, Indiana; he obtained an average reduction of the surface of 1.5 mm in 43 years or 0.14 inch (3.5 mm) in 100 years. Professor A. Geikie, in the late 1800s, calculated a rate of lowering of about a third of an inch (8.5 mm) per century for marble tombstones in another humid region: Edinburgh, Scotland. In 1983, in subhumid Grand Forks, North Dakota, I arrived at a conservative maximum rate of about 1 mm in 97 years or 0.04 inch (1.0 mm) in 100 years for the lower marble part of a marble-granite tombstone (Figures 23–2, 23–3). Conservative, I say, because measurements were taken to the crests of narrow ridges presumably caused by foliation, and they, themselves, were dissolved. Keep in mind that the rate of solution depends on the surface measured (lower versus upper) and direction of exposure. More solution tends to occur on the side from which most rains arrive, especially if accompanied by wind.

Physical weathering of marble is evidenced by cracking and flaking. Much of this results from frost wedging in higher latitudes. Southern exposures of tombstones tend to show most cracking and flaking because of more frequent freezing and thawing. Some flaking is also likely due to the expansive force of newly formed salts that occupy greater volume. In a polluted, sulfur dioxide atmosphere, sulfuric acid reacts with marble to form calcium sulfate or gypsum. This newly formed gypsum—with about a twofold increase in volume—may force apart calcite crystals and flake the marble.

Granite tombstones are much more resistant to weathering than those of marble. Flaking, though, may be obvious and perhaps iron oxide staining from chemical weathering of the ferromagnesian minerals—again, depending on the climate and the time

FIGURE 23–2. Comparison of weathering effects of two rocks in a 97-year-old tombstone (1886 to 1983) in subhumid Grand Forks, northeastern North Dakota. The lower, inscribed part of fine-grained marble shows obvious effects of solution; the fine, diagonal ridges and crack (left) presumably reflect foliation of the marble. The upper granite pillar is essentially unweathered except for barely discernible iron oxide staining—not evident in the photograph.

FIGURE 23–3. Close-up of the tombstone in Figure 23–2 where the granite (upper) overlies the marble (lower). The marble is cracked, lichen-covered, and displays low oblique ridges at the top brought out by solution.

involved. Surprisingly, perhaps, some have found that polished granite weathers faster than unpolished granite. Professor Geikie saw no apparent sign of decay of granite in the Edinburgh churchyards after 20 years. I saw only barely evident tinges of iron oxide staining on polished granite in the Grand Forks example (Figures 23–2, 23–3). An extreme case of resistance is the granite of monuments along the Nile in arid Egypt dating back to 2850 B.C. In 1916, D. C. Barton, an American geologist, estimated the average rate of disintegration as only about 1 to 2 mm in 1000 years, which equates to about 0.006 inch in 100 years; the maximum rate was only about 0.02 inch (0.5 mm) per 100 years.

Of the other rocks apt to be used as tombstones, the weathering effects of limestone are similar to those of marble. Sandstone may be extremely durable if the grains are largely quartz and cemented by silica. Look for weathering in sandstone tombstones where grains are held together by soluble carbonate minerals or iron oxides. G. F. Matthias, a New york high school instructor, examined tombstones—in humid Middletown, Connecticut—of reddish brown feldspar-quartz sandstone ("brownstone," see Chapter 22) cemented by calcite and hematite. He measured chemical weathering, in the 1960s, by using the depth of inscriptions, based on an assumed standard depth of 0.16 inch on essentially unweathered slate tombstones. Result: an average rate of weathering of about 0.03 inch in 145 years or 0.02 inch (0.6 mm) in 100 years. In nearby West Willington, Connecticut, Professor P. H. Rahn found, in the 1960s, that this sandstone weathers faster than marble, followed by schist and granite. Accessory structures in sedimentary rocks such as fossils, concretions, or nodules (see Chapter 16) may result in accelerated weathering. And if the rock is well bedded and cut so the beds are oriented upward, weathering attack may be intense if other resistive qualities are lacking. Slate is particularly resistant to weathering because, usually being a metamorphosed shale, it consists mostly of clay minerals derived already by weathering, especially of feldspars, and is essentially stable. You are quite apt to see, within a humid climate, 200-year-old slate tombstones, perfectly legible, in the same graveyard with younger, illegible marble tombstones.

Since we are all mortal, why not apply our knowledge of tombstone weathering to personal practical use? Choosing a tombstone for yourself or others, you realize now that a marble or limestone tombstone just doesn't have the "permanence" of one of granite, silica-cemented sandstone, or slate, especially in a humid climate. But the choice depends, too, on cost and attractiveness to the beholder.

Selected Readings

GEIKIE, ARCHIBALD. "Rock-Weathering, as Illustrated in Edinburgh Churchyards," *Proceedings of the Royal Society of Edinburgh,* 10 (1880), 518–532.

MATTHIAS, G. F. "Weathering Rates of Portland Arkose Tombstones," *Journal of Geological Education,* 15 (1967), 140–144.

24 HOW TO PROSPECT FOR GOLD

Prospecting for useful earth materials follows a similar approach whether the desired material is uranium, copper, petroleum, coal, water—or gold. You must be able to (1) recognize the material, perhaps in several forms; (2) be aware of its occurrence, especially with regard to rock types; and (3) acquire and operate the necessary equipment (or have someone else do so) for its extraction. In prospecting, science is certainly foremost, with its careful observations and systematizing and analyzing of facts, but art is also important, art in the sense of having a feeling where discovery may occur and the predisposition to try a theory or supposition.

I'll focus on gold to illustrate the prospecting process for three reasons: (1) it still excites the thrill of discovery in most of us; (2) in our day of sophisticated equipment and big business, gold—as a useful earth material—can still be sought by the individual with limited funds while on the move; and (3) any gold discovered can be sold readily, again by the individual.

Recognizing Gold

As almost everyone knows, gold is bright, yellow, glitters, and is heavy. So that's all there is to recognizing it—right? Not quite. A gold- or sun-yellow is characteristic of gold in its native or chemically uncombined state, but gold may be pale yellow to silver-white if the combined silver—with which it is usually alloyed—is appreciable. Gold glitters because of its metallic luster, but so do many other metallic minerals. The metallic fool's gold minerals, pyrite and chalcopyrite (see Chapter 14) are brittle and yield a blackish powder (streak) when scratched. Gold, on the other hand, is malleable—readily pounded into thin sheets—and displays its own color when easily scratched by a knife. Iron-stained biotite mica flakes are often confused for flakes of gold. The mica flakes are readily crushed with a knife or needle, and float away with water currents. Gold is heavy; there is no disputing that. Expect native gold in rock as feathery leaves, wires, or plates; in sediment it occurs as flakes, grains, or larger nuggets. Gold dissolves readily in mercury to form an amalgam or alloy, utilized as a tooth filling. Only one acid—a mixture of three parts hydrochloric acid and one part nitric acid—will dissolve gold.

Another commercial solvent, sodium or calcium cyanide, dissolves gold. It is the cyanide process that is used to free and concentrate finely disseminated gold from pulverized rock on a commercial scale.

Which brings us to another point. Gold may be present and not visible. Such is the case of gold combining chemically with tellurium to form the tellurides. Or gold may be so finely disseminated in gold or other ore that it cannot be seen. Although obviously unrecognizable, such gold can be mined profitably if processed appropriately.

Where To Look

Gold must originate at considerable depth, carried upward by hot fluids—from a magma—that force their way into rock fractures. Crystallization, largely in quartz veins, occurs as the fluids cool and pressures decrease. So the ultimate source of rock-borne or **lode gold** is igneous rocks. But just any igneous rocks won't do. The most favorable ones are light-colored, fine-grained rocks or felsites (see Chapter 15) in the form of flows, sills, and dikes (see Figure 15–14), and light-colored porphyries. Borders of granitic masses may be productive. Other rocks, igneous, metamorphic, or sedimentary, tend to contain little gold unless transected by felsite rock bodies. Your best bet, though, is to search quartz veins. But lode gold is not commonly visible. That doesn't mean, though, that you should forego searching for it. Examining abandoned mine dumps for lode gold may give you a good feeling for suitable rock associations.

As lode deposits break up by weathering, such materials are carried downslope and usually accumulate as stream sediment. **Placer gold** is found here, the most ready source for the lightly-equipped traveler. Gold, being heavier than most materials, concentrates naturally in crevices, low spots, behind boulders, and on bars—any place where stream flow is checked. Irregular bedrock surfaces on stream floors, such as foliation surfaces of slate and schist, act as natural riffles to trap gold. Look especially in streams where both erosion and deposition (see Chapter 2) are going on. Examine, too, stream terraces now being cut into: They represent former stream deposits. Where possible, follow gold-bearing streams to a lake or the sea. Here, along beaches, wave action winnows sand and sorts and concentrates gold and other heavy minerals (see Figure 4–3).

Now, to zero in on the search. Look, especially, where gold has been found or mined previously. Most of the gold retrieved in the United States has come from only five states: Alaska, California, Colorado, Nevada, and South Dakota. But many more have produced gold. Check with J. E. Ransom's *The Gold Hunter's Field*

Book for hundreds of gold-producing localities in 32 states as well as the Canadian provinces. Another, perhaps less readily available, source is the U. S. Geological Survey's Professional Paper 610, *Principal Gold-Producing Districts of the United States.* Use topographic maps (see Chapter 1) to help find suitable localities and abandoned mines, and use geologic maps (see Chapter 20) to locate suitable rock types. Generally, unreserved federal land—such as that for parks and military purposes—and much state land is open to prospecting. National forests are especially likely places; detailed maps of these are readily available from ranger stations and similar outlets. Always make a reasonable attempt to gain permission to prospect on any private land; the lack of such permission may lose any subsequent access for you or other traveling prospectors. Respect, too, the other ethics of collecting covered in Chapter 21.

Panning

Panning is the accepted method for an individual to extract placer gold on a small scale. The shallow, steel gold pan is 15 to 18 inches (38 to 46 cm) in diameter with a broad, low-sloping brim. You can usually pick up one from a hardware store in mining areas. Lacking a gold pan, try a frying pan from which any grease has been removed. A large wooden or plastic salad bowl will do in a pinch. A long-handled, round-pointed shovel is necessary, and a pick or mattock may prove useful in loosening sediment or soil. Have your hand lens handy for spotting any fine gold traces in the pan, and acquire tweezers and vials or small bottles for picking and storing gold flakes or grains.

Now for the actual panning. To become efficient faster, watch a panner first if possible. Lacking this, begin this way. Fill the pan about half full with stream sand and gravel, using your shovel. Submerge the pan in water, break up any earth clumps, and pick out the coarse gravel particles. Shake the pan in a circular motion, and tilt one edge away from you to allow the lighter materials to wash away. Slow down as a heavy mineral concentrate begins to accumulate in the bottom of the pan. Much of the concentrate tends to be dark gray or black, consisting of such minerals as magnetite, platinum, and stream tin. Magnetite, especially common, is black, metallic, brittle, and attracted by a magnet. Platinum, more valuable than gold, is gray-white, metallic, malleable like gold, and may be attracted by a magnet. Stream tin, actually cassiterite, is gray, greasy-appearing, and brittle. Examine the concentrate with your hand lens, and pick out any gold with tweezers.

If water is scarce, fill a tub full and pan there. And if water is unavailable, you might try a dry-wash or winnowing process. Pick out by hand, or better yet, screen out the coarser particles. Place the fine fraction on a blanket and have your partner help you shake it in a wind. The gold and other heavy minerals remain on the blanket as the finer and lighter particles are blown away.

One last point about panning or winnowing. I've emphasized

placer gold because it is generally more readily visible. If, however, you strongly suspect gold in a well-weathered lode deposit, by all means try panning or winnowing the loose, weathered material.

Selling (Any of) Your Gold

Unusual gold in rock, including large flakes, wires, leaves, crystals, and large or unusual nuggets, is usually sold as specimen gold to museums or private collectors. The price is negotiable and depends on what a buyer is ready and willing to pay. Other gold, that in the finer sizes, usually must be assayed before sale because gold is never 100 percent pure. For very small amounts, the assay costs could consume most of your profit. In mining areas, you are apt to find several private buyers. Inquire about them to help ensure locating one that is fair and trustworthy. You might also wish to sell through banks, which can be relied upon and maintain good records of transactions. Whatever the outlet, keep a close eye on world gold prices—fluctuating daily—to assure yourself of a fair price.

Selected Readings

KOSCHMANN, A. H., and M. H. BERGENDAHL. "Principal Gold-Producing Districts of the United States," *U. S. Geological Survey Professional Paper* 610 (1968), 1–283.

PEARL, R. M. *Handbook For Prospectors.* New York: McGraw-Hill Book Company, 1973.

RANSON, J. E. *The Gold Hunter's Field Book.* New York: Harper and Row, Publishers, Inc., 1975.

WEST, J. M. "How to Mine and Prospect For Placer Gold," *U. S. Bureau of Mines Information Circular* 8517 (1971), 1–43.

Appendixes

Appendix A: Museums with Geological Materials in the United States and Canada, Exclusive of Those in Parks and Similar Places

United States

ALABAMA

Anniston: Anniston Museum of Natural History, The John B. Lagarde Environmental Interpretive Center
Birmingham: Red Mountain Museum
University: University of Alabama Museum of Natural History

ALASKA

Central: Circle District Historical Society, Incorporated, Museum
Fairbanks: University of Alaska Museum
Juneau: Alaska Historical Library and Museum
Nome: Carrie M. Mclain Memorial Museum

ARIZONA

Bisbee: Bisbee Civic Center and Mining and Historical Museum
Flagstaff: Museum of Astrogeology, Meteor Crater
Flagstaff: Museum of Northern Arizona
Page: Carl Hayden Visitor Center
Phoenix: Arizona Mineral Resources Museum
Phoenix: The Arizona Museum
Phoenix: Pueblo Grande Museum
Sedona: American Meteorite Museum
Tempe: Center for Meteorite Studies
Tucson: Arizona-Sonoro Desert Museum
Tucson: University of Arizona Mineral Museum

ARKANSAS

Arkadelphia: Henderson State University Museum
Fayetteville: University of Arkansas Museum
Jonesboro: Arkansas State University Museum
Smackover: Oil Heritage Center

CALIFORNIA

Berkeley: Museum of Paleontology, University of California
Carlsbad: San Luis Rey Historical Society
Claremont: Raymond M. Alf Museum
Fullerton: Museum of North Orange County
Independence: Eastern California Museum
Los Angeles: Geology Museum and Collections, University of California
Los Angeles: Natural History Museum of Los Angeles County
Oakland: The Oakland Museum
Pacific Grove: Pacific Grove Museum of Natural History
Palm Springs: Palm Springs Desert Museum, Incorporated
Randsburg: Desert Museum
Redlands: San Bernadino County Museum
Riverside: Jurupa Mountains Cultural Center
Riverside: Riverside Municipal Museum
Sacramento: Sacramento Science Center and Junior Museum
San Diego: Allison Center for the Study of Paleontology
San Diego: Natural History Museum
San Francisco: California Division of Mines and Geology Mineral Museum
San Francisco: Josephine D. Randall Junior Museum
San Jacinto: San Jacinto Valley Museum
San Jose: Youth Science Institute
Santa Barbara: Santa Barbara Museum of Natural History
Walnut Creek: Alexander Linsay Junior Museum
Yucaipa: Mousley Museum of Natural History
Yucca Valley: Hi-Desert Nature Museum
Yreka: Siskiyou County Museum

COLORADO

Bayfield: The Gem Village Museum
Boulder: University of Colorado Museum
Canon City: Canon City Municipal Museum
Colorado Springs: May Natural History Museum
Cripple Creek: Cripple Creek District Museum, Incorporated
Denver: Denver Museum of Natural History
Golden: Colorado School of Mines, Geology Museum
Grand Junction: Museum of Western Colorado
Gunnison: Gunnison County Pioneer and Historical Society
Idaho Springs: Clear Creek Historic Mining and Milling Museum
Saguache: Saguache County Museum
Sterling: Overland Trail Museum
Telluride: San Miguel Historical Society
Trinidad: Trinidad State Junior College Museum

CONNECTICUT

New Fairfield: Hidden Valley Nature Center
New Haven: Peabody Museum of Natural History

New London: Thames Science Center, Incorporated
Stamford: Stamford Museum and Nature Center
West Hartford: Children's Museum of Hartford
Westport: Nature Center for Environmental Activities

DELAWARE

Newark: University of Delaware Museum
Wilmington: Society of Natural History of Delaware Museum

DISTRICT OF COLUMBIA

Explorers Hall
National Museum of Natural History
Rock Creek Nature Center
United States Department of Interior Museum

FLORIDA

Bradenton: South Florida Museum and Bishop Planetarium
Daytona Beach: Museum of Arts and Sciences
DeLand: Gillespie Museum of Minerals, Stetson University
Fort Lauderdale: Discovery Center
Fort Pierce: St. Lucie County Historical Museum
Gainesville: Florida State Museum, University of Florida
Jacksonville: Jacksonville Museum of Arts and Sciences
Miami: Museum of Science
Orlando: The John Young Science Center
Ormond Beach: Tomoko Museum
Pensacola: T. T. Wentworth, Jr. Museum
Safety Harbor: Safety Harbor Museum of History and Fine Arts, Incorporated
St. Petersburg: The Science Center of Pinellas County
West Palm Beach: Science Museum and Planetarium of Palm Beach County, Incorporated

GEORGIA

Athens: Georgia State Museum of Science and Industry
Athens: University of Georgia Museum of Natural History
Atlanta: Fernbank Science Center
Atlanta: Parks, Recreation and Historical Sites Division, Georgia Department of Natural Resources
Augusta: Augusta Richmond County Museum
Buford: Lanier Museum of Natural History
Columbus: Columbus Museum of Arts and Sciences, Incorporated
Elberton: Elberton Granite Museum and Exhibit
Elberton: Elberton Granite Museum and Exhibit

HAWAII

Hilo: Lyman House Memorial Museum
Honolulu: Bernice Pauahi Bishop Museum

IDAHO

Moscow: College of Mines, University of Idaho
Pocatello: Idaho Museum of Natural History
Sandpoint: Bonner County Historical Museum
Wallace: Coeur D'Alene District Mining Museum
Weiser: Intermountain Cultural Center and Museum

ILLINOIS

Carbondale: University Museum
Chicago: Chicago Academy of Sciences, Museum of Ecology
Chicago: Field Museum of Natural History
Chicago: Museum of Science and Industry
Elmhurst: Lizzadro Museum of Lapidary Art
Normal: University Museums, Illinois State University
Quincy: Quincy and Adams County Museum
Rock Island: Fryxell Geology Museum
Rockford: Burpee Museum of Natural History
Springfield: Illinois State Museum
Urbana: Museum of Natural History, University of Illinois

INDIANA

Evansville: Evansville Museum of Arts and Science
Indianapolis: The Children's Museum
Indianapolis: Indiana State Museum
New Harmony: Historic New Harmony, Incorporated
Richmond: Joseph Moore Museum

IOWA

Cedar Falls: University of Northern Iowa Museum
Cherokee: Sanford Museum and Planetarium
Davenport: Putnam Museum
Des Moines: Des Moines Center of Science and Industry
Des Moines: Iowa State Historical Department, Division of Historical Museum and Archives
Iowa City: Museum of Natural History
Iowa Falls: Ellsworth College Museum
Sioux City: Sioux City Public Museum
Waterloo: Grout Museum of History and Science

KANSAS

Baldwin: Baker University Museum
Hays: Fort Hays State University Museums
Hays: Sternberg Memorial Museum
La Crosse: Post Rock Museum
Lawrence: Systematics Museums, University of Kansas
Manhattan: Kansas State University, Fairchild Hall
McPherson: McPherson Museum
Oakley: Fick Fossil and History Museum

KENTUCKY

Berea: Berea College Museums
Louisville: Museum of History and Science
Owensboro: Owensboro Area Museum
Wickliffe: Ancient Buried City

LOUISIANA

Baton Rouge: Louisiana State University Museum of Geoscience
Shreveport: Grindstone Bluff Museum and Environmental Education Center

MAINE

Augusta: Maine State Museum
Caribou: Nylander Museum
Castine: Wilson Museum
Hinckley: L. C. Bates Museum
Lewiston: Department of Geology, Bates College

MARYLAND

Baltimore: Cylburn Museum
Baltimore: Maryland Academy of Sciences
Solomons: Calvert Marine Museum

MASSACHUSETTS

Amherst: The Pratt Museum of Natural History
Boston: Museum of Science
Cambridge: Mineralogical Museum of Harvard University
Lincoln: Drumlin Farm Educational Center
Milton: Blue Hills Trailside Museum
Pittsfield: The Berkshire Museum
Springfield: Springfield Science Museum
Worcester: Worcester Science Center

MICHIGAN

Alpena: Jesse Besser Museum
Ann Arbor: The University of Michigan Exhibit Museum
Battle Creek: Kingman Museum of Natural History
Bloomfield Hills: Cranbrook Institute of Science
East Lansing: The Museum, Michigan State University
Grand Rapids: Grand Rapids Public Museum
Houghton: A. E. Seaman Mineralogical Museum
Lansing: Carl G. Fenner Arboretum and Environmental Education Center
Mount Pleasant: Center for Cultural and Natural History
Ossineke: Dinosaur Gardens, Incorporated
Traverse City: Great Lakes Area Paleontological Museum

MINNESOTA

Granite Falls: Yellow Medicine County Historical Museum
Minneapolis: James Ford Bell Museum of Natural History
Redwing: Goodhue County Historical Society
St. Paul: The Science Museum of Minnesota
Walker: Walker Museum

MISSISSIPPI

Flora: Mississippi Petrified Forest
Hattiesburg: John Martin Frazier Museum of Natural Science
Jackson: Mississippi Museum of Natural Science
State College: Dunn-Seiler Museum, Mississippi State University

MISSOURI

Chesterfield: Babler Nature Interpretive Center
Columbia: University of Missouri, Columbia, Department of Geology
Fayette: Stephens Museum of Natural History and The United Methodist Museum of Missouri
Joplin: Tri-State Mineral Museum
Kansas City: Kansas City Museum
Lebanon: Nature Interpretive Center
Rolla: Ed Clark Museum of Missouri Geology
Rolla: University of Missouri, Rolla, Geology Museum
St. Joseph: St. Joseph Museum
St. Louis: Museum of Science and Natural History

MONTANA

Bozeman: Museum of the Rockies
Butte: Mineral Museum
Ekalaka: Carter County Museum
Fort Peck: Fort Peck Museum
Havre: Northern Montana College Collections
Lewistown: Central Montana Museum
Loma: Earth Science Museum
Roundup: Musselshell Valley Historical Museum
Superior: Mineral County Museum and Historical Society

NEBRASKA

Agate: Cook Museum of Natural History
Chadron: Chadron State College Earth Science Museum
Hastings: Hastings Museum
Kearney: Fort Kearney Museum
Lincoln: University of Nebraska State Museum

NEVADA

Carson City: The Nevada State Museum
Henderson: Clark County Southern Nevada Museum

Las Vegas: Museum of Natural History, University of Nevada
Reno: Mackay School of Mines Museum

NEW HAMPSHIRE

Dover: Annie E. Woodman Institute
Durham: Geology Museum, University of New Hampshire
Hanover: Montshire Museum of Science
Suncook: Bear Brook Nature Center

NEW JERSEY

Franklin: Franklin Mineral Museum
Morristown: Morris Museum of Arts and Sciences
Mountainside: Trailside Nature and Science Center
Newark: Newark Museum
Paramus: Bergen Community Museum
Paterson: Paterson Museum
Princeton: Princeton University Museum of Natural History
Rutherford: Meadowlands Museum

NEW MEXICO

Abiquiu: Ghost Ranch Living Museum
Albuquerque: Institute of Meteoritics Meteorite Museum
Los Alamos: Los Alamos County Historical Museum
Portales: Miles Museum
Socorro: New Mexico Bureau of Mines Mineral Museum

NEW YORK

Albany: New York State Museum
Bear Mountain: Bear Mountain Trailside Museums
Buffalo: Buffalo Museum of Science
Cornwall-on-Hudson: Museum of The Hudson Highlands
Cross River: Trailside Nature Museum
Hicksville: The Gregory Museum: Long Island Earth Science Center
Ithaca: Paleontological Research Institution
Mount Kisco: Westmoreland Sanctuary, Incorporated
New York: American Museum of Natural History
Niagara Falls: Schoellkopf Geological Museum
Pawling: Museum of Natural History
Richfield Springs: Petrified Creatures Museum of Natural History
Rochester: Rochester Museum and Science Center
Schenectady: Museum of the Department of Geology, Union College
Staten Island: Staten Island Institute of Arts and Sciences
Stony Brook: Museum of Long Island Natural Sciences
Syosset: Nassau County Museum, Division of Museum Services and Department of Recreation and Parks
Utica: Children's Museum of History, Natural History and Science

NORTH CAROLINA

Asheville: Colburn Memorial Mineral Museum
Chapel Hill: Research Laboratory of Anthropology and Museum of Geology, University of North Carolina
Charlotte: Charlotte Nature Museum
Charlotte: Discovery Place
Durham: North Carolina Museum of Life and Science
Gastonia: Schiele Museum of Natural History and Planetarium, Incorporated
Greensboro: The Natural Science Center of Greensboro, Incorporated
Hickory: Catawba Science Center
High Point: Environmental Education, Recreation and Research Center, Incorporated
Lake Waccamaw: Lake Waccamaw Depot Museum
Raleigh: North Carolina State Museum
Salisbury: Dan Nicholas Park Nature Center
Spruce Pine: Museum of North Carolina Minerals
Winston-Salem: Nature Science Center

NORTH DAKOTA

Epping: Buffalo Trails Museum
Grand Forks: Department of Geology Museum, University of North Dakota

OHIO

Antwerp: Ehrhart Museum
Bay Village: Lake Erie Nature and Science Center
Cincinnati: Cincinnati Museum of Natural History
Cincinnati: Trailside Nature Center and Museum
Cincinnati: University of Cincinnati Geology Museum
Cleveland: Cleveland Museum of Natural History
Columbus: Center of Science and Industry
Columbus: Ohio Historical Center
Columbus: Orton Geological Museum, Ohio State University
Dayton: Dayton Museum of Natural History
Glenford: Flint Ridge Memorial Museum
Lima: Allen County Museum
Norwalk: Firelands Historical Society Museum
Tiffin: Jones Collection of Minerals and Biology Museum
Toledo: Toledo Museum of Health and Natural History

OKLAHOMA

Alva: Northwestern Oklahoma State University Museum
Duncan: Stephens County Historical Museum
Norman: Stovall Museum of Science and History
Stillwater: Oklahoma State University Museum of Natural and Cultural History
Tonkawa: The A. D. Buck Museum of Natural History and Science

OREGON

Ashland: Oregon Museum of Natural History
Eugene: Natural History Museum
Portland: Oregon Museum of Science and Industry
Portland: State of Oregon Department of Geology and Mineral Industries Museum
The Dalles: The Nichols Museum
Tillamook: Tillamook County Pioneer Museum

PENNSYLVANIA

Ashland: Ashland Anthracite Museum
Bryn Mawr: Bryn Mawr College, Department of Geology
Harrisburg: Pennsylvania Historical and Museum Commission
Philadelphia: Academy of Natural Sciences of Philadelphia
Philadelphia: Franklin Institute
Philadelphia: Wagner Free Institute of Science
Pittsburgh: Carnegie Museum of Natural History, Carnegie Institute
Scranton: Everhart Museum of Natural History, Science and Art
Titusville: Drake Well Museum
West Chester: West Chester State College Museum
Wilkes-Barre: Wyoming Historical and Geological Society

RHODE ISLAND

Providence: Roger Williams Park Museum

SOUTH CAROLINA

Charleston: Charleston Museum
Greenwood: The Museum
Rock Hill: Museum of York County
Spartanburg: Spartanburg County Nature-Science Center

SOUTH DAKOTA

Kadoka: Badlands Petrified Gardens
Piedmont: Black Hills Petrified Forest
Rapid City: Museum of Geology, South Dakota School of Mines and Technology

TENNESSEE

Knoxville: Geological Museum, University of Tennessee
Memphis: Southwestern at Memphis
Nashville: Cumberland Museum and Science Center

TEXAS

Austin: Texas Memorial Museum
Big Bend: Alpine's Museum
Brazosport: Brazosport Museum of Natural Science
Bryan: Brazos Valley Museum of Natural Science
Canyon: Panhandle-Plains Historical Museum

College Station: Texas A and M College Museum
Dallas: Dallas Museum of Natural History
Dallas: Southwest Museum of Science and Technology, The Science Place
El Paso: El Paso Centennial Museum, University of Texas at El Paso
Fort Worth: Fort Worth Museum of Science and History
Forth Worth: The Western Company Museum
Houston: Environmental Science Center
Houston: Houston Museum of Natural Science
Houston: NASA Manned Spacecraft Center
Iraan: Iraan Museum
Lubbock: The Museum, Texas Tech University
McKinney: Heard Natural Science Museum and Wildlife Sanctuary, Incorporated
Midland: Midland County Historical Museum
Midland: Permian Basin Petroleum Museum, Library, and Hall of Fame
Seguin: Fiedler Memorial Museum
Waco: John K. Strecker Museum

UTAH

Lehi: John Hutchings Museum of Natural History
Price: Prehistoric Museum of the College of Eastern Utah
Salt Lake City: Utah Museum of Natural History

VERMONT

Burlington: Robert Hull Fleming Museum, University of Vermont
St. Johnsbury: Fairbanks Museum of Natural Science

VIRGINIA

Blacksburg: Museum of the Geological Sciences
Charlottesville: Brooks Museum, University of Virginia
Harrisonburg: D. Ralph Hostetter Museum of Natural History
Newport News: Peninsula Nature and Science Center
Richmond: Science Museum of Virginia
Roanoke: Roanoke Valley Science Museum

WASHINGTON

Port Angeles: Pioneer Memorial Museum
Seattle: Pacific Science Center
Seattle: Thomas Burke Memorial Washington State Museum
Spokane: Eastern Washington State Historical Society Museum

WEST VIRGINIA

Charleston: Sunrise Foundation, Incorporated
Huntington: Geology Museum

WISCONSIN

Campbellsport: Henry S. Reuss Ice Age Visitor Center
Green Bay: Neville Public Museum of Brown County
La Crosse: Viterbo College Museum
Madison: Geology Museum, Science Hall, University of Wisconsin
Madison: University of Wisconsin Zoological Museum
Milwaukee: Green Memorial Museum, University of Wisconsin-Milwaukee
Milwaukee: Milwaukee Public Museum
New London: New London Public Museum
Platteville: Platteville Mining Museum
Stevens Point: The Museum of Natural History

WYOMING

Buffalo: Johnson County, Jim Gatchell Memorial Museum
Greybull: Greybull Museum
Jackson: Jackson Hole Museum Foundation
Lander: Pioneer Museum
Laramie: The Geological Museum, The University of Wyoming

Canada

ALBERTA

Calgary: Glenbow Museum
Drumheller: Drumheller and District Museum Society
Edmonton: Department of Geology Museum
Edmonton: Provincial Museum of Alberta
Medicine Hat: Medicine Hat Museum and Art Gallery

BRITISH COLUMBIA

Kamloops: Kamloops Museum and Archives
Skidegate: Queen Charlotte Islands Museum
Vancouver: British Columbia and Yukon Chamber of Mines
Vancouver: Geological Museum, University of British Columbia
Vancouver: M. Y. Williams Geological Museum
Victoria: British Columbia Mineral Museum

MANITOBA

Brandon: B. J. Hales Museum of Natural History
Winnipeg: Manitoba Museum of Man and Nature

NEW BRUNSWICK

St. John: The New Brunswick Museum

NEWFOUNDLAND

St. John's: Memorial University of Newfoundland, Department of Geology

NOVA SCOTIA

Parrsboro: Parrsboro Geological Mineral and Gem Museum

ONTARIO

Cobalt: Cobalt Northern Ontario Mining Museum
Kingston: Miller Museum of Geology and Mineralogy
Oil Springs: Oil Museum of Canada
Ottawa: National Museum of Natural Sciences
Toronto: Ontario Science Center
Toronto: Royal Ontario Museum

QUEBEC

Montreal: Redpath Museum of Mcgill University
Quebec: Musée De Géologie Et De Minéralogie

SASKATCHEWAN

Regina: Museum of Natural History
Swift Current: Swift Current Museum

Appendix B: United States State Geological Surveys

Geological Survey of Alabama
Drawer O, University, AL 35486

Alaska Division of Geological &
Geophysical Surveys
Pouch 7-028, Anchorage, AK 99501

Arizona Bureau of Geology &
Mineral Technology
845 North Park Avenue
Tucson, AZ 85719

Arkansas Geological Commission
Vardelle Parham Geology Center
3815 West Roosevelt Road
Little Rock, AR 72204

California Division of Mines &
Geology
1416 Ninth Street, Room 1341
Sacramento, CA 95814

Colorado Geological Survey
1313 Sherman Street, Room 715
Denver, CO 80203

Connecticut Geological & Natural
History Survey
Room 553, State Office Building
165 Capitol Avenue
Hartford, CT 06115

Delaware Geological Survey
University of Delaware
101 Penny Hall
Newark, DE 19711

Florida Bureau of Geology
903 West Tennessee Street
Tallahassee, FL 32304

Georgia Geologic Survey
Department of Natural Resources
Room 400, 19 Martin Luther King Jr.
Drive SW
Atlanta, GA 80334

Hawaii Division of Water & Land
Development
Box 373, Honolulu, HI 96809

Idaho Bureau of Mines & Geology
Morrill Hall, Room 332
University of Idaho
Moscow, ID 83843

Illinois Geological Survey
Natural Resources Building
615 East Peabody Drive
Champaign, IL 61820

Indiana Geological Survey
611 North Walnut Grove
Bloomington, IN 47405

Iowa Geological Survey
123 North Capitol Street
Iowa City, IA 52242

Kansas Geological Survey
1930 Avenue A, Campus West
University of Kansas
Lawrence, KS 66044

Kentucky Geological Survey
311 Breckinridge Hall
University of Kentucky
Lexington, KY 40506

Louisiana Geological Survey
Box G, University Station
Baton Rouge, LA 70893

Maine Geological Survey
Department of Conservation
State House, Station 22
Augusta, ME 04333

Maryland Geological Survey
The Rotunda, 711 West 40th Street,
Suite 440
Baltimore, MD 21211

Massachusetts Department of
Environmental Quality Engineering
Division of Waterways, 1 Winter
Street
7th Floor
Boston, MA 02108

Michigan Geological Survey
Division
Box 30028
Lansing, MI 48909

Minnesota Geological Survey
1633 Eustis Street
St. Paul, MN 55108

Mississippi Bureau of Geology
Box 5348
Jackson, MS 39216

Missouri Division of Geology &
Land Survey
Box 250
Rolla, MO 65401

Montana Bureau of Mines &
Geology
Montana College of Mineral Science
& Technology
Butte, MT 59701

Nebraska Conservation & Survey
Division
113 Nebraska Hall
University of Nebraska
Lincoln, NE 68588

Nevada Bureau of Mines & Geology
University of Nevada
Reno, NV 89557

New Hampshire Department of
Resources & Economic
Development
117 James Hall
University of New Hampshire
Durham, NH 03824

New Jersey Geological Survey
CN-029
Trenton, NJ 08625

New Mexico Bureau of Mines &
Mineral Resources
Campus Station
Socorro, NM 87801

New York State Geological Survey
3136 Cultural Education Center
Empire State Plaza
Albany, NY 12230

North Carolina Department of
Natural Resources & Community
Development
Geological Survey Section
Box 27687
Raleigh, NC 27611

North Dakota Geological Survey
University Station
Grand Forks, ND 58202

Ohio Division of Geological Survey
Fountain Square, Building B
Columbus, OH 43224

Oklahoma Geological Survey
830 Van Vleet Oval, Room 163
Norman, OK 73019

Oregon Department of Geology &
Mineral Industries
1005 State Office Building
1400 SW Fifth Avenue
Portland, OR 97201

Pennsylvania Bureau of Topographic & Geologic Survey
Box 2357
Harrisburg, PA 17120

Rhode Island Statewide Planning
Program
265 Melrose Street
Providence, RI 02907

South Carolina Geological Survey
Harbison Forest Road
Columbia, SC 29210

South Dakota Geological Survey
Science Center
University of South Dakota
Vermillion, SD 57069

Tennessee Division of Geology
701 Broadway
Nashville, TN 37203

Texas Bureau of Economic Geology
University of Texas
Box X, University Station
Austin, TX 78712

Utah Geological & Mineral Survey
606 Black Hawk Way
Salt Lake City, UT 84108

Vermont Geological Survey
Agency of Environmental
Conservation
Heritage II Office Building
Montpelier, VT 05602

Virginia Division of Mineral
Resources
Box 3667
Charlottesville, VA 22903

Washington Division of Geology & Earth Resources
Department of Natural Resources
Olympia, WA 98504

West Virginia Geological & Economic Survey
Box 879
Morgantown, WV 26507

Wisconsin Geological & Natural History Survey
1815 University Avenue
Madison, WI 53706

Wyoming Geological Survey
Box 3008, University Station
University of Wyoming
Laramie, WY 82071

Appendix C: Regional Geological Highway Maps of the American Association of Petroleum Geologists[1]

ALASKA AND HAWAII

GREAT LAKES REGION
Illinois
Indiana
Michigan
Ohio
Wisconsin

MID-ATLANTIC REGION
Delaware
Kentucky
Maryland
North Carolina
South Carolina
Tennessee
Virginia
West Virginia

MID-CONTINENT REGION
Arkansas
Kansas
Missouri
Oklahoma

NORTHEASTERN REGION
Connecticut
Maine
Massachusetts
New Hampshire

[1]Most geological maps also include cross sections, a tectonic map, a landform map, and geological history of a region. Some maps provide mineral, rock, and fossil collecting localities and places of geological interest. Order regional maps from: The American Association of Petroleum Geologists, P. O. Box 979, Tulsa, OK 74101.

New Jersey
New York
Pennsylvania
Rhode Island
Vermont

NORTHERN
GREAT PLAINS REGION

Iowa
Minnesota
Nebraska
North Dakota
South Dakota

NORTHERN
ROCKY MOUNTAIN REGION

Idaho
Montana
Wyoming

PACIFIC NORTHWEST REGION

Idaho (in part)
Oregon
Washington

PACIFIC SOUTHWEST REGION

California
Nevada

SOUTHEASTERN REGION

Alabama
Florida
Georgia
Louisiana
Mississippi

SOUTHERN
ROCKY MOUNTAIN REGION

Arizona
Colorado
New Mexico
Utah

TEXAS

INDEX

A **boldfaced** page number indicates where a term is defined or explained; an *italicized* page number indicates an illustration. Where a feature is both defined or explained and illustrated on the same page, a **boldfaced** number is used.

Q

R

S

T